DATE DUE	

Atlas of
VISUALIZATION
III

Edited by
The Visualization Society of Japan

CRC Press
Boca Raton   New York

Acquiring Editor: Joel Claypool
Contact Editor: Ben Kato
Associate Editor: Felicia Shapiro
Project Editor: Jennifer Richardson
Marketing Manager: Susie Carlisle
Direct Mail Marketing Manager: Becky McEldowney
Prepress: Gary Bennett
Cover Designer: Denise Craig
Manufacturing: Sheri Schwartz

Library of Congress Cataloging-in-Publication Data

Cataloging and ISSN number available from the Library of Congress.

This book contains information obtained from authentic and highly regarded sources. Reprinted material is quoted with permission, and sources are indicated. A wide variety of references are listed. Reasonable efforts have been made to publish reliable data and information, but the author and the publisher cannot assume responsibility for the validity of all materials or for the consequences of their use.

Neither this book nor any part may be reproduced or transmitted in any form or by any means, electronic or mechanical, including photocopying, microfilming, and recording, or by any information storage or retrieval system, without prior permission in writing from the publisher.

All rights reserved. Authorization to photocopy items for internal or personal use, or the personal or internal use of specific clients, may be granted by CRC Press LLC, provided that $.50 per page photocopied is paid directly to Copyright Clearance Center, 27 Congress Street, Salem, MA 01970 USA. The fee code for users of the Transactional Reporting Service is ISBN 0-8493-2656-7/94/$0.00+$.50. The fee is subject to change without notice. For organizations that have been granted a photocopy license by the CCC, a separate system of payment has been arranged.

CRC Press LLC's consent does not extend to copying for general distribution, for promotion, for creating new works, or for resale. Specific permission must be obtained in writing from CRC Press LLC for such copying.

Direct all inquiries to CRC Press LLC 2000 Corporate Blvd., N.W., Boca Raton, Florida 33431.

© 1997 by CRC Press LLC

No claim to original U.S. Government works
Printed in the United States of America 1 2 3 4 5 6 7 8 9 0
Printed on acid-free paper

Preface

Visualization is a new science where invisible information is made positively visible using visualization techniques and computers. Through such visible information new information is obtained which helps to clarify the phenomena.

To promote this science, the Visualization Society of Japan (VSJ) was established as an official society registered in the Government of Japan in 1990. This society is the developed version of the Flow Visualization Symposium which started in 1973.

The Atlas of Visualization was first published as the English Journal of the VSJ in 1992. In view of our experience preparing our first issue, the Editorial Committee was reorganized so that major countries are cordially solicited to have Editors in order to facilitate the committee management. Under this new system, the Atlas of Visualization Volume II was published in May 1996.

Now the VSJ has decided to publish the Atlas of Visualization Volume 3 with the same policy as the first two.

The book consists of 16 papers and 12 gravures that were recommended from the Regional Editors of respective countries. The contents are roughly divided into physical visualization and computer aided visualization. Their subjects cover vortex, turbulent flow, PIV, PTV, mass transfer, heat transfer, water spray, jet flames, combustion, medical science, thermal plume, and numerical simulation.

The biggest feature of this book is full color pages where the complex phenomena can be expressed more clearly and understood more correctly.

Lastly, on behalf of the Editorial Committee, we would like to give our sincere acknowledgment to the authors who presented their valuable papers and photographs to this series for exchanging new knowledge through it.

Yasuki Nakayama
Yoshimichi Tanida
Editors-in-Chief

      Y. Nakayama   **Y. Tanida**

Editorial Board

Editors-in-Chief

Y. Nakayama	(Future Technology Research Institute, Japan)
Y. Tanida	(Tokai University, Japan)

Regional Editors

R. J. Adrian	(University of Illinois, USA)
R. F. Boucher	(University of Manchester, U.K.)
G. M. Carlomagno	(University of Naples, Italy)
C. J. Chen	(Florida State University, USA)
Y. N. Chen	(National Taiwan University, Taiwan)
K. C. Cheng	(University of Alberta, Canada)
P. Hébrard	(ENSICA, France)
N. Kasagi	(University of Tokyo, Japan)
T. Kobayashi	(University of Tokyo, Japan)
V. Kottke	(Hohenheim University, Germany)
M. G. Mungal	(Stanford University, USA)
F. Ogino	(Kyoto University, Japan)
M. Ohashi	(Chuo University, Japan)
V. Tésar	(CVUT-Technical University, Czech R.)
Q. D. Wei	(Peking University, China)
T. Yoshida	(National Aerospace Laboratory, Japan) [Secretary of Editorial Board]

Advisory Editors

F. Durst	(University of Erlangen-Nürnberg, Germany)
L. M. Fingerson	(TSI, USA)
N. Fomin	(University of Minsk, Belarus)
J. P. Gostelow	(University of Leicester, U.K.)
S. K. Graffunder	(Cray Research Inc., USA)
E. I. Hayes	(DANTEC, Denmark)
L. Hesselink	(Stanford University, USA)
B. Khalighi	(GMRL, USA)
J. Kim	(EPRI, USA)
R. Kimura	(University of Tokyo, Japan)

G. E. A. MEIER	(M P I Göttingen, Germany)
S. MURAKAMI	(University of Tokyo, Japan)
H. M. NAGIB	(Illinois Institute of Technology, USA)
H. NAKAGAWA	(Kyoto University, Japan)
S. ROBINSON	(NASA Langley, USA)
K. TAKAYAMA	(Tohoku University, Japan)
T. TANAHASHI	(Keio University, Japan)
K. TSUCHIYA	(Waseda University, Japan)
V. R. WATSON	(NASA Ames, USA)
M. YANO	(NEWJEC, Inc., Japan)
E. F. ZHIGALKO	(St. Petersburg University, Russia)

Advisory Board

T. ANDO	(Keio University, Japan)
T. ASANUMA	(University of Tokyo, Japan)
O. DRACOS	(ETH, Switzerland)
R. J. GOLDSTEIN	(University of Minnesota, USA)
T. HAYASHI	(Chuo University, Japan)
M. HINO	(Chuo University, Japan)
I. IMAI	(University of Tokyo, Japan)
S. J. KLINE	(Stanford University, USA)
T. S. LEE	(Seoul University, Korea)
H. OHASHI	(Kogakuin University, Japan)
S. TANEDA	(Kurume Institute of Technology, Japan)
F. J. WEINBERG	(Imperial College, U.K.)
J. ZIEREP	(University of Karlsruhe, Germany)

Contents

Chapter One
Quantitative Visualization of Three-Dimensional Free Surface Slopes and Elevations 1
Dana Dabiri, Xin Zhang, and Morteza Gharib

Chapter Two
Dynamic and Multiple Laser Light Sheets: An Experimental Approach of Unstationary and Three-Dimensional Flows 23
J. P. Prenel, R. Porcar, A. Texier, and A. Strzclecki

Chapter Three
Some Typical Mechanisms in the Early Phase of the Vortex-Shedding Process from Particle-Streak Visualization 43
Madeleine Coutanceau and Gerard Pineau

Chapter Four
Visualization of Velocity and Vorticity Fields 69
Anjaneyulu Krothapalli, Luiz Lourenco, and Chiang Shih

Chapter Five
Colors in PIV 83
A. Cenedese and G. P. Romano

Chapter Six
Flow Visualization in Turbulent Large-Scale Structure Research 99
Lorenz W. Sigurdson

Chapter Seven
Analysis of Turbulence and Vortex Structures by Flow Mapping 115
C. A. Greated, C. E. Damm, and J. Whale

Chapter Eight
Quantitative Visualization of 2-D and 3-D Flows Using a Color-Coded Particle Tracking Velocimetry 131
Tzong-Shyan Wung

Chapter Nine
Visualization and Determination of Local Mass Transfer at Permeable and Nonpermeable Walls in Liquid Flow 143
W. Kühnel and V. Kottke

Chapter Ten
Heat Transfer to Air from a Yawed Circular Cylinder 153
Gennaro Cardone, Guido Buresti, and Giovanni Maria Carlomagno

Chapter Eleven
A Visualization Study on Water Spray of Dragon Washbasin 169
Qing-Ding Wei, Da-Jun Wang, Xiang-Dong Du, and Jun Chen

Chapter Twelve
Dynamics of Propane Jet Diffusion Flames 181
V. R. Katta, L. P. Goss, W. M. Roquemore, and L.-D. Chen

Chapter Thirteen
Application of Visualization Techniques for Studying the Internal Combustion Engine 199
Satoshi Yamazaki and Akinori Saitoh

Chapter Fourteen
3D and 4D Visualization of Morphological and Functional Information from the Human Body Using Noninvasive Measurement Data 213
Naoki Suzuki and Akihiro Takatsu

Chapter Fifteen
Giant Thermal Plume Generation Over Asphalt-Paved Highway 225
Manabu Kanda and Mikio Hino

Chapter Sixteen
Numerical Flow Visualization of Posuk-Chung, the Ninth Century Remains of Poetry Making Curved Water Channel in Korea 241
Keun-Shik Chang and Eunbo Shim

Index 251

Frontispiece Illustrations

Figure 1   R. D. Keane and R. J. Adrian
University of Illinois at Urbana-Champaign

Figure 2   W. H. Finlay and Y. Guo
University of Alberta

Figure 3   M. G. Mungal[1] and A. Lozano[2]
Stanford University[1] and LITEC/CSIC[2]

Figure 4   T. Sugano
Kyoto University

Figure 5   N. Suzuki
Jikei University

Figure 6   K. Takayama, H. Babinsky, and J. M. Yang
Tohoku University

Figure 7   V. R. Katta[1], L. P. Goss[1], W. M. Roquemore[2], and L.-D. Chen[3]
Innovative Scientific Solutions, Inc.[1], Wright Laboratory, WL/POSC,[2] and the University of Iowa[3]

Figure 8   Wallops Flight Facility
NASA

Figure 9   F. P. Welsh and K. Warburton
Defense Research Agency

Figure 10   T. Kobayashi and T. Kogaki
University of Tokyo

Figure 11   K. Kinoshita
Kagoshima University

Figure 12   I. Nezu
Kyoto University

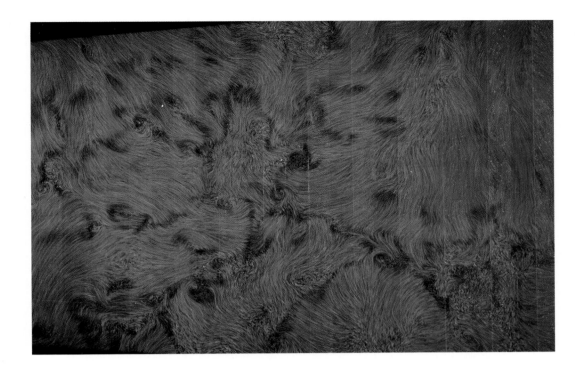

PLAN-FORM VIEW OF THE TEMPERATURE FIELD OF UNSTEADY TURBULENT NON-PENETRATIVE THERMAL CONVECTION IN THE VICINITY OF A GROOVED ROUGH SURFACE

Figure 1 Planform flow visualization of non-penetrative thermal convection 5 mm above a rough surface plate with step heights (amplitude) of 5.08 mm and step widths of 12.7 mm (wavelength of 25.4 mm) in a layer depth of 100 mm. Encapsulated thermochroic crystals are suspended in the water to illustrate the structure of the geometry of cells of downward flowing, colder fluid and their interaction with the roughness elements and the buoyant plumes for which $Ra_f = 2.59 \times 10^9$. Kodak Gold 400 photographic film with $f^\# = 4$ and an exposure time of 2 seconds enables particle streaks to be seen.

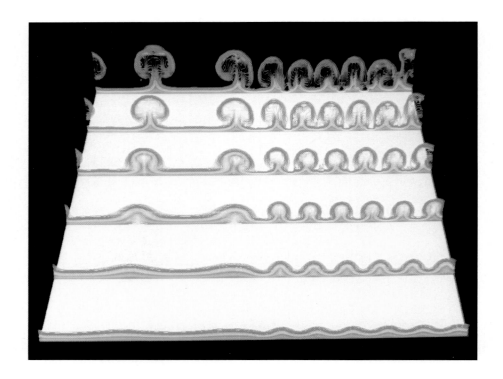

GÖRTLER VORTICES DEVELOPING IN A CONCAVE BOUNDARY LAYER

Figure 2 Contours of streamwise velocity are shown in cross-flow planes for a spatially developing simulation of Görtler vortices obtained from a spectral element simulation. The viewer is looking downstream from outside the boundary layer. The characteristic mushroom-shaped patterns of the vortex pairs become distorted by a spanwise secondary instability in which the Görtler vortices are unstable to perturbations with spanwise wavelengths different than their own. This instability largely causes irregularity in Görtler vortex patterns, but in extreme cases can cause pairs of vortices to merge together and can also cause the formation of new pairs of vortices between existing pairs.

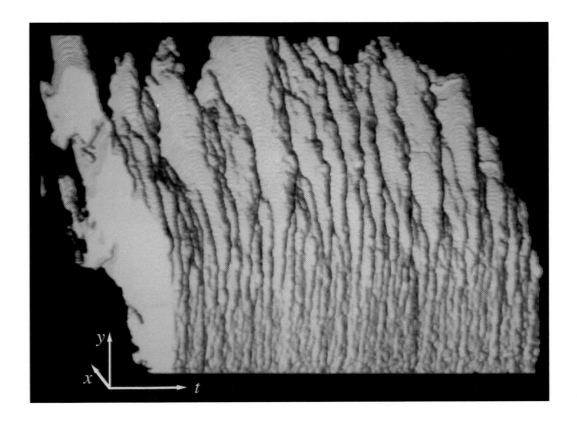

VOLUME RENDERING OF A LARGE BURNING JET IN CROSSFLOW

Figure 3 The photograph shows a volume-rendering of 300 sequential images of a large oil well discharge burning in an ambient crosswind. The images were first recorded as a movie sequence of a burning jet in crossflow. Each movie frame is then digitized to yield an x-y image, with sequential images similarly digitized as a function of time, t. The images are then stacked onto each other, on an image computer, to produce an x-y-t volume rendering. The rendered object is illuminated from the left by a light source, with hidden surfaces removed, and can be viewed from any desired orientation. For the rendering shown here, the leftmost face shows the first x-y jet image onto which the subsequent images are stacked to the right. The rendering shows that the jet consists of a set of organized structures (the travelling bands) which merge into larger structures as they progress downstream; the structures appear quasi-periodic, with final burnout at the flame tip. These phenomena have been observed for free jet flames as well. See also "Some observations of a large, buring jet in crossflow," M.G. Mungal and A. Lozano, *Experiments in Fluids*, 21 (1996), 264–267.

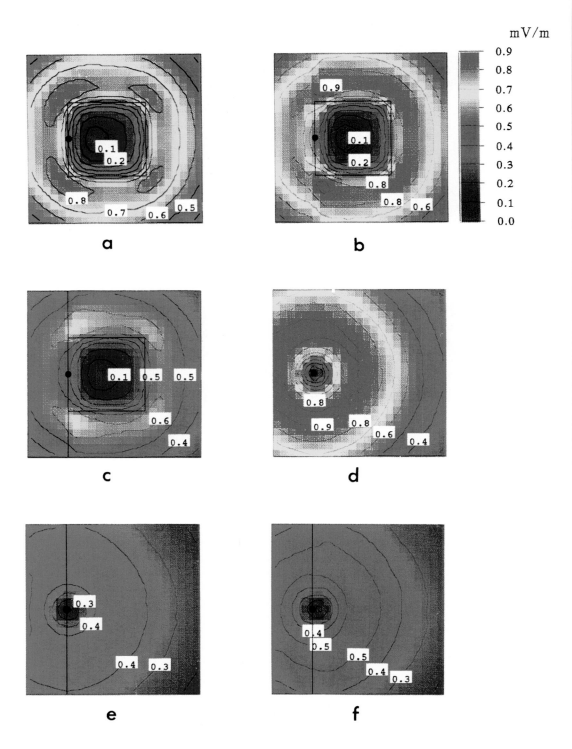

ELECTRIC FIELD VISUALIZATION FOR EARTH RESOURCE AND ENVIRONMENTAL SYSTEM PREDICTIVE DESIGN

Figure 4 Visualization of subsurface information by the electric field response (mV/m) due to (a) a cubic conductive target with the resistivity of 5 Ω · m located at the depth of 20 m in the homogeneous medium of 100 Ω · m, (b) the cubic inhomogeneity under the horizontal layer of 300 Ω · m, (c) the target with the vertical fault of 50 Ω · m, (d) the effect of near-surface horizontal layer, (e) the effect of vertical fault, and (f) the effects of a horizontal layer and a vertical fault of a 3-D model in the case of the subsurface current source of 90 m depth and the surface potential electrodes, that is the subsurface-to-surface solid method. The electric field has become a new sensitive imaging information aid to the detailed investigation or the improved monitoring for the earth resource and environmental system predictive design. (Refer to Journal of the Visualization Society of Japan, Vol. 15. No. 58, pp. 196–205, 1995).

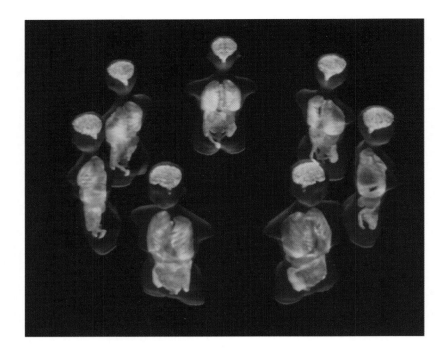

3D Image of a Female Reconstructed with Serial MRI Data

Figure 5 It is an ideal of medicine to be able to observe the invisible internal structures of a human. Noninvasive measurement techniques, developed for diagnostic imaging, and computer graphics techniques have allowed us to realize this ideal. The picture shows 3D images of the whole female body of a living human, reconstructed for a 3D human atlas in a medical data base.
Serial MRI images are utilized for the 3D reconstruction of the body structure. The shape and location of the principal organs in the thoracic and abdominal cavities are visible. Organ textures are mapped on 3D organ surfaces to give informtion about the surface anatomy.

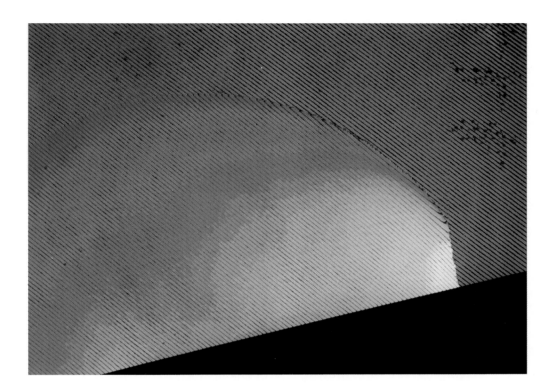

Shock Wave Reflection over a Wedge

Figure 6 This is a finite fringe holographic interferogram of shock wave reflection from a 15 degree wedge for shock wave Mach number 1.2 at atmospheric air. The experiment was carried out in a 60 mm × 150 mm diaphragmless shock tube equipped with double exposure holographic interferometry. The interferometric image was stored on a computer by an image scanner and the fringes were then automatically analyzed by Fourier fringe analysis. The fringe shift was determined with the accuracy of one tenth of the fringe interval and the interpolated fringeshifts were displayed with false colors. The false color display was superimposed with the original finite fringe interferogram.

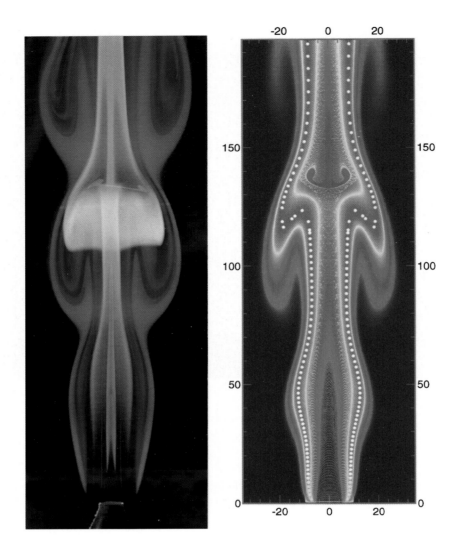

STRUCTURES OF METHANE JET DIFFUSION FLAMES

Figure 7 Structures of low-speed laminar and moderate-speed transitional methane jet diffusion flames are shown in Figures a and b, respectively. In each figure, instantaneous images of the laboratory flame obtained using Reactive-Mie-Scattering (RMS) technique and the computed flame obtained by superimposing particle traces over temperature field are shown on the left half and right half, respectively. Direct numerical simulations of these flames are carried out by solving time-dependent Navier-Stokes equations along with species and energy conservation equations on a staggered cylindrical coordinate system. A detailed chemical kinetics model that consists of seventeen species and involving in 52 elementary reactions is used. Strong buoyancy-induced toroidal vortices located outside the flame surface are dominating both the flows. The Kelvin-Helmholz vortices developed in the transitional flame are slowly dissipating due to the higher viscosity and buoyancy force associated with the high-temperature combustion products.

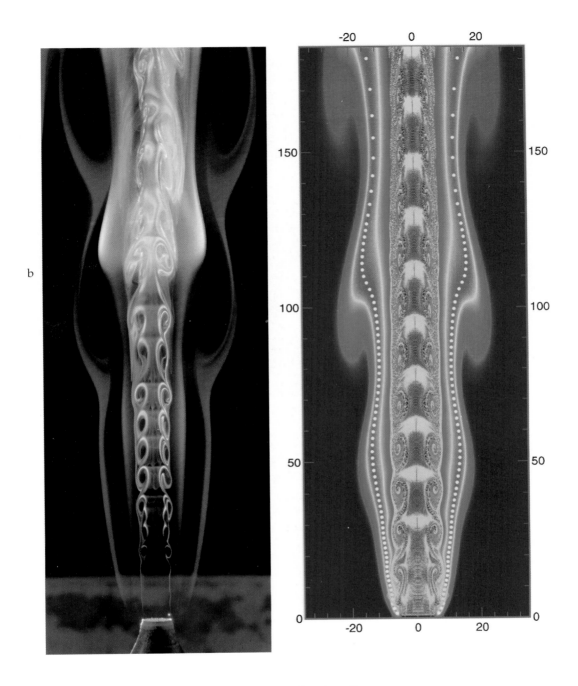

Figure 7 (continued).

WINGTIP VORTEX

Figure 8 Picture of Thrus crop-duster airplane flying over a red smoke source, thus visualizing the left wingtip vortex. This experiment was done at the U.S. NASA (National Aeronautics and Space Administration) Wallops Flight Facility, approximately thirty five years ago. (Illustration suggested by A. Roshko, California Institute of Technology).

a

b

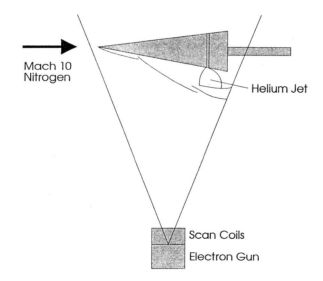

MACH 10 FLOW OVER A CONE WITH JET

Figure 9 The details of the interaction between a Mach 10 flow over a cone and a sonic helium jet issuing from a slot near its base are made visible by scanning an electron beam to excite a plane of fluorescence light. The intensity of the light is directly related to the gas density, while the color depends upon the species, nitrogen appearing violet and helium green. The design of the model allows direct comparison to axisymmetric computational fluid dynamics simulations.

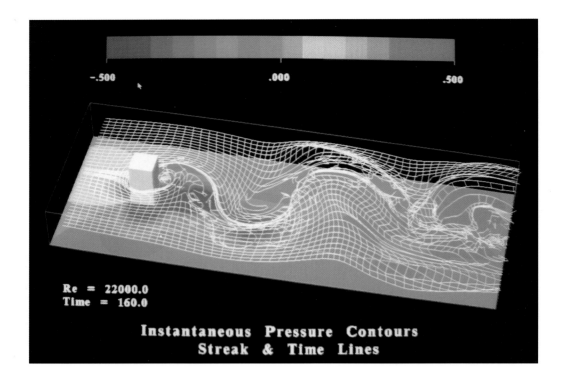

Turbulent Flow Behind a Rectangular Cylinder by LES

Figure 10 This picture shows the streak lines, the time lines, and the pressure distribution of a turbulent flow around a stationary rectangular cylinder at high Reynolds number of RE = 22,000. This is calculated by large eddy simulation (LES) in which only large eddies are directly simulated and small scale eddies are modeled using a subgrid scale (SGS) model. The formation of the Kármán vortex street with fluctuations is clearly observed and the flow behind the rectangular cylinder is fully three-dimensional in this high Reynolds number.

LANDSAT IMAGE OF A VOLCANIC CLOUD FROM MT. SAKURAJIMA

Figure 11 This aerial color image of Kagoshima prefecture is reproduced from multispectral sensor (MSS) data taken by satellite Landsat-5 with bands 4, 5, and 7 (0.5–0.6, 0.6–0.7, and 0.8–1.1 μm) as blue, red, and green. The satellite scan was taken at 10:08 on March 13, 1990 over southern Kyushu, Japan, when the volcano Sakurajima was continuously ejecting a cloud of volcanic ash, water vapor, and other gases such as sulfur dioxide. The crater height is 1040 m above the sea level, and the 100 km cloud floated at an altitude of 1500 m in the free atmosphere where NNW winds were blowing at around 8 m/s. Similar patterns of linear dispersion from the volcano are often seen in satellite images, as well as other patterns such as fan- and belt-type dispersion.

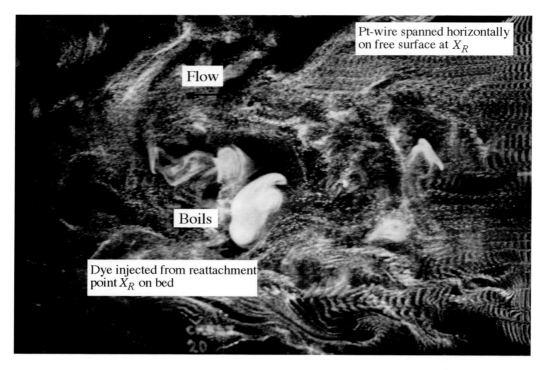

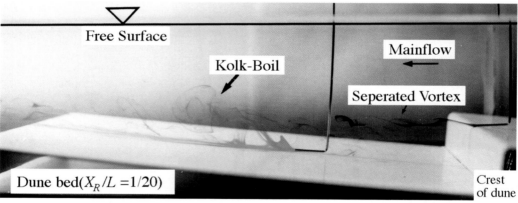

Coherent Vortices behind a Dume in Open-Channel Flows

Figure 12 Coherent vortices such as separated vortex and kolk-boil are generated behind dunes in rivers. It is a very important topic in hydraulic and river engineering to investigate such coherent structures because kolk-boils cause suspended sediment from sandbeds up to the water surface. Figure a shows a complicated interaction between the separated vortices and large-scale boils behind the dunes. Separated vortices generate kolk-boils at the reattachment point that is fluctuated in low-frequency. Figure b shows the water surface behind the dune by simultaneous use of hydrogen-bubbles very near the water surface and dyes injected from the reattachment point. Low-speed cauliflower-like boils are seen clearly and are associated with the dye.

chapter one

Quantitative Visualization of Three-Dimensional Free Surface Slopes and Elevations

Dana Dabiri, Xin Zhang, and Morteza Gharib

Graduate Aeronautics Laboratory, California Institute of Technology, Pasadena, California

Abstract—*A new technique that integrates optics, colorimetry, and digital image processing is developed to measure the three-dimensional surface slope and elevation for time-evolving flows. The setup requires that a lens be placed at a distance of its focal length away from a diffuse, uniform, white light source which is covered with an appropriate transparent color palette. This setup then produces a system of color rays where beams of the same color are parallel to themselves, yet oriented in different directions with respect to beams of other colors. This arrangement is used in one of two modes. In the first mode, the reflective mode, the arrangement is set to illuminate the free surface from above the free surface. A three-chip color camera placed far above the free surface captures the* **reflected** *rays from the free surface (hence the name* **reflective** *mode). In the second mode, the refractive mode, the arrangement illuminates the free surface from below the free surface. A three-chip color camera placed far above the free surface captures the* **refracted** *rays from the free surface (hence the name* **refractive** *mode). For both modes, each of the three RGB color signals from the color camera is recorded onto separate laser disk recorders. These modes then allow for measurements of the free surface slope by creating a one-to-one correspondence between different colors and the free surface slopes. These data can then be integrated to derive the free surface elevation. Applications to various problems are discussed.*

Introduction

Recently, there has been an increased interest in the behavior of free surface phenomena. One such interest lies in understanding short surface wind waves because of their importance in mass, momentum, and energy exchanges at the air–sea interface, microwave remote sensing of a sea surface, generation of underwater near-surface ambient sound, and also in understanding the theory of wave–wave interactions. For these wind waves, the air–sea free surface interfacial area varies significantly as the wave steepness varies. Also, the wind affects the roughness of the free surface, and therefore influences the development of wind waves. Due to these complex issues, the shape of very finite height, short surface waves are still not clearly understood and experimental observation is needed.

Another area of interest is in understanding turbulence near the free surface, since it exhibits behavior that is different from that found within the fluid bulk. Three-dimensional (3-D) homogenous turbulence within the fluid bulk implies a $-5/3$ slope of the energy spectra, where the energy flows from the large-scale structures to the small-scale structures. However, for a shear-free surface, the kinematic boundary condition requires that vortex filaments terminate at a normal angle with the free surface (Lugt, 1987). This means that the terminating vortices can only possess a normal component to the free surface. For the ideal case of a flat surface, the vortices at the surface can be treated as two-dimensional (2-D) vortices where all the vortex filaments are parallel. For high Reynolds number flows, the condition of two-dimensionality requires conservation of enstrophy which in combination with the conservation of energy imposes a reverse cascade (Kraichnan, 1971). This reverse energy cascade results in the formation of large vortical structures at the free surface, which in turn creates free surface deformations. Therefore, a central issue in understanding the free surface turbulence is to relate the surface elevation to the near-surface flow field.

In this respect, the lack of a global surface mapping technique which could reveal the temporal evolution of the surface elevation has prevented the progress of viable research in areas such as these. As such, techniques have been developed to measure free surface slopes and elevations. With these techniques which have been developed, it is undesirable to use material probes extending through the water surface when studying small surface structures due to the interference between the probe and the fluid to be studied. We will therefore review optical methods that have been used to measure a small structure of fluid surface.

Most of the early attempts were based on stereo photography (Kohlshutter, 1906; Laas, 1905). The method (Holthuijsen, 1983; Banner et al., 1989) can also be used to measure short waves, but these are unreliable. The principle of this method is to calculate the surface elevations from the displacement between corresponding points in two images obtained from two different viewpoints. However, measuring a water surface that is specular and curved provides severe problems of mismatched points (Jahne et al., 1994; Zhang and Cox, 1994).

Measurement of statistical properties of ocean surface slopes on a large area from light reflection on the water surface has a long history (Schooley, 1954; Cox and Munk, 1954). Longuet-Higgins (1960) presented a mathematical analysis of light reflection and refraction from a rippled surface. Photographs have been used by Barber (1949; Stilwell, 1969, Kasevich, 1975; Lubard et al., 1980; Ebuchi, 1987) to infer the 2-D wave spectrum.

An accurate way to measure short wind waves is to measure surface slope by refraction of light at the air–water interface (Cox, 1958). As pointed out by Cox (1958), it is more convenient to use the slope of the water surface rather than the more usual elevation from mean water level as the parameter to be studied, because high-frequency waves, although of low height, have large slopes. Also, the energy of the pure capillary waves is proportional to the mean square of wave slope. The optical technique has been extended by Wu (1969), who used a point source of light to measure not only the slope, but the curvature as well. Later, a laser light source was introduced to measure wave slopes (Lubard et al., 1980; Tober et al., 1973). Keller and Gotwols (1983), Jahne and Waas (1989), and Jahne and Riemer (1990) have developed a method of measuring 2-D wave slopes one slope component at a time. A video camera was used in these recent 2-D slope detractors. Zhang and Cox (1994) recently developed a color light encoding technique for recording surface gradients (both orthogonal slope components) of an area of water with color photography. 2-D water surface elevations can then be completely recovered from the recorded gradient data. This technique has been successfully applied to measure short wind waves in a wave tank (Zhang and Cox, 1994; Zhang, 1995). Dabiri et al. (1994) and Zhang et al. (1994) have extended this work to allow for real-time data acquisition with the use of a 3-CCD color

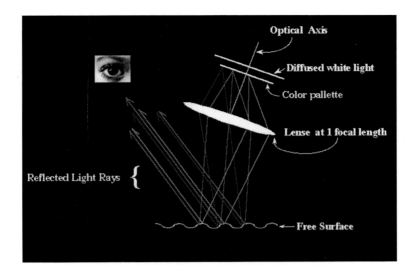

Figure 1 Optical principle demonstrating the setup for the reflective mode.

camera. This work showed measurements of near-free surface turbulence interacting with and deforming the free surface.

In this paper, we would like to present a review of the 2-D free surface color mapping technique. First, fundamental elements, principles, and discussion of errors given in Zhang and Cox (1994) are discussed for conciseness and clarity. Then applications to various problems are demonstrated.

Optical Principle for Measuring Free Surface Slopes

The basic principle for measuring free surface slopes is to color code the different slopes with different colors. This, then, establishes a one-to-one correspondence between color and slope, thus making free surface slope measurements possible. There are two variations on the principle of the technique developed for measuring both components of the free surface slope. The first is the *reflective mode*, where light is reflected from the free surface toward the data acquisition camera. The second is the *refractive mode*, where light is refracted through the free surface toward the data acquisition camera. Detailed descriptions of the two principles are given below.

Reflective Mode

The idea of this mode is illustrated in Figure 1. A uniform white light source covered with an appropriate transparent color palette is placed far below the surface. Sample palettes used by Zhang and Cox (1994) are shown in Figures 6 to 8. This allows different colors to emanate from the light source depending on the color's position within the palette. A lens is placed above the light source at its focal length's distance from the light source. Thus, each set of color light rays emanating from the light source onto the lens will emerge from the lens parallel to themselves, but oriented in different directions with respect to other color beams. For example, if red is positioned in the center of the light source, all red rays passing through the lens will emerge parallel to themselves and in the direction of the optical axis. If blue is positioned to the right of the light source, all blue rays emerging from the lens would be parallel to themselves, but slanted to the left with respect to the optical axis. Likewise, if green is positioned to the left of the light source, all green rays emerging from the lens would be parallel to themselves, yet slanted to the right with respect to the

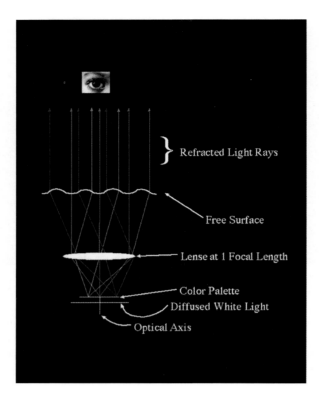

Figure 2 Optical principle demonstrating the setup for the refractive mode.

optical axis. When this system of parallel color beams is used to illuminate a free surface from above the surface, the color rays will reflect from the free surface. A 3-CCD color camera located far away from the free surface captures the reflected beams. Since the camera is located far away from the free surface, only almost-parallel beams enter the camera. The error associated with nonparallel beams entering the camera will be discussed in the section "Data Analysis". Thus, due to the local free surface slope, only the rays that are reflected by the free surface toward the camera are captured by the camera. This procedure guarantees a one-to-one correspondence between the color of the beam reflected into the camera and the local free surface slope. Thus, this approach color codes the different slopes of the free surface with different colors.

Refractive Mode

The approach is very similar to that described in the previous section. The idea of the method is illustrated by Figure 2. As can be seen in Figure 2, a lens, color palette, and light source are set up in an arrangement identical to that described in the previous section to create a system of color beams where all beams of the same color are parallel to themselves, yet oriented in different directions with respect to beams of other colors. When this system of parallel color beams is used to illuminate a free surface from below the surface, the color rays will refract from the free surface. For acquisition, Zhang and Cox (1994) used a photographic camera. As in the previous section, it is recommended to use a 3-CCD color camera located far away from the free surface to capture the refracted beams. Thus, due to the local free surface slope, only the rays that are refracted by the free surface toward the camera are captured by the camera. This guarantees a one-to-one correspondence between the color of the beam refracted into the camera and the local free surface slope, and therefore color codes the different slopes of the free surface with different colors.

Properties of Both Modes

Even though the two modes are slightly different, there are elements that are common to both and should be mentioned:

1. Although throughout this paper we demonstrate and apply this technique to water free-surfaces, the technique will work for any clear fluid. Moreover, for the reflective case, the fluid need not even be fully transparent. The only requirement is that the fluid must be transparent enough so that the color light beams are not scattered appreciably or too greatly absorbed.
2. The water surface area to be examined is limited to the area where *all* the color beams illuminate the free surface. This is determined by the devices used to generate the color light beams such as the lens and the color palette. This will be discussed further in the section "Data Analysis".
3. It is important to make sure that the color-to-slope calibration and measurements be such that they be independent of position. To ensure this, it is necessary to place the camera far from the examined area so that all color rays entering the camera are almost parallel. If the camera is too close to the free surface, the same color rays will enter the camera from different angles with respect to the free surface, thus making the color-to-slope measurements a function of position.

System Design

In the previous section, "Optical Principle for Measuring Free Surface Slopes", the water slope detection system is described in principle for both the reflective and refractive modes. In this section, we will discuss the system design, the calibration procedure, and the error analysis in detail for both the reflective and the refractive modes. The maximum area that can be measured, L_{max}, depends on the size of the lens, D, the maximum possible angle of the color light with respect to the optical axis, β_{max}, the distance along the optical axis between the lens and the free surface, H, and the angle between the optical axis and the free surface normal, β_0, according to the following equation:

$$L_{max} = D\cos\beta_0 - \left(H - \frac{D}{2}\sin\beta_0\right)\tan\beta_{max} - \left(H + \frac{D}{2}\sin\beta_0\right)\tan(\beta_{max} - 2\beta_0) \quad (1)$$

The maximum angle of the water surface slope, $\theta_{1\,max}$, that can be measured is determined by the size of the color palette relative to the focal length of the lens. For the refractive mode, this relationship can be expressed as:

$$\sin(\theta_{1\,max}) = n \cdot \sin(\theta_{1\,max} - \theta_0) \quad (2)$$

$$\sin(\theta_0) = \frac{1}{n} \cdot \sin(\beta_{max}) \quad (3)$$

where n is the optical refraction index of the water relative to air (Figure 3). For the reflective mode, the maximum angle of the water surface slope, $\theta_{1\,max}$, that can be measured is equal to the incident angle of the color rays onto the surface:

$$\theta_{1\,max} = \theta_{0\,max} \quad (4)$$

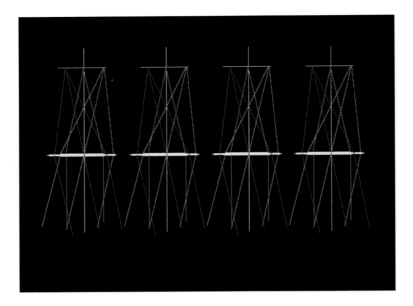

Figure 3 The optical setup involving multiple arrays of lenses and color palettes. This allows for viewing large areas while maintaining large F #s per lens.

Typically, in the refracted mode, the reflected light angle is ~1/3 of the incident color light beam angle, whereas for the reflected mode, the reflected light angle is identical to the incident color beam angle. Thus, the refracted mode is better suited for free surface flows with large angles, while the reflected mode is better suited for free surface flows with smaller angles. Zhang and Cox (1994) have found that the reflective mode is better suited for measuring angles of up to ~10°, while the refractive mode is capable of measuring angles up to ~50°. Likewise, aside from the range of angles to be measured, it is important to know what is the smallest change of angle that can be measured. For the reflective mode, this has been found to be ~1°, while for the refractive mode, the smallest detectable change in angle is ~5°.

Furthermore, a large value of β_{max} tends to make the optical characteristics of the lens far from ideal. It is therefore recommended to use lenses with large F numbers. Unfortunately, as the F number of the lens gets larger, the maximum measurable area becomes smaller. Thus, in order to circumvent this problem, several color beam generating devices can be placed side by side, thus making it possible to measure larger areas of the free surface (see Figure 4). Also, for applications where it is necessary to place the color beam generating system below the free surface, it is recommended to place this system at least at about half of the dominant long wavelength in order to avoid influencing the water flow of these waves by the bulk of the submerged lens system.

Once it has been decided which of the two modes is to be used, the final setup is placed around the experiment. Figure 5 shows a typical setup. The camera is connected to a timing unit, which then strobes the flashlamp. The timing unit also synchronizes the real-time image acquisition with the Sony LVR/LVS laser disk recorders. Note that each of the three color signals (R, G, and B) is recorded onto three separate units. Color NTSC video signal has a bandwidth of about 7.8 MHz. The chrominance signal, which carries information regarding the color content, is compressed within this bandwidth. Therefore, the color resolution is not very good. To circumvent this problem, colors emanating from the free surface are acquired and recorded with a color video signal that is composed of three individual NTSC signals. This increases the bandwidth of the acquired images by a factor of 3 to about 24 MHz, which therefore provides proper image resolution. After the real-time image acquisition, images are captured with a PC/AT color frame grabber and

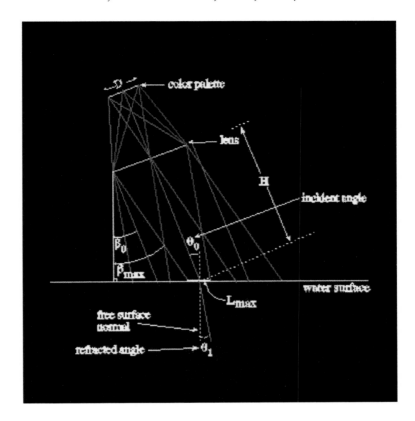

Figure 4 Ray tracing geometry demonstrating the maximum measurable area on the free surface.

analyzed in order to extract the free surface slope content. It should be mentioned that even though Figure 5 shows the setup for the reflective mode, the acquisition setup still holds for the refractive mode as well.

It was briefly mentioned in "Optical Principle for Measuring Free Surface Slopes" that the data acquisition color camera was placed far from the free surface in order to ensure

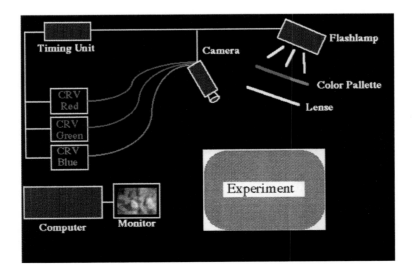

Figure 5 Block diagram showing the experimental setups.

almost parallel beams entering the camera. This can now be quantified. For the reflective mode, given that the size of each color element is ΔCE, the slope sensitivity, which is defined as the minimum slope change that can be measured, is

$$\Delta \alpha \leq tan^{-1} (\Delta CE/f) \tag{5}$$

where f is the local length of the lens. The viewing angle of the camera is also given as

$$\psi = tan^{-1} (D/TL) \tag{6}$$

where D is the diameter of the camera lens and TL is the total length from the free surface to the camera. Therefore, in order to prevent any superpositioning of the different colors, the viewing angle of the camera must be less than the slope sensitivity, or:

$$\frac{D}{TL} < \frac{\Delta CE}{f} \tag{7}$$

Therefore, the camera must be placed at a certain distance from the free surface given by:

$$TL > \frac{Df}{\Delta CE} \tag{8}$$

For the refractive mode, the above relations hold when ΔCE is replaced by $\Delta CE/n$, where n is the index of refraction.

This setup is then used for performing both a calibration and the experiment. For calibration, a surface with a known curvature is placed within the measurement area. The system of parallel color beams is then used to illuminate this surface, and images are then acquired. Since the curvature of the surface is known, the colors can then be easily calibrated to the slope by following the procedure outlined in the section "Data Analysis".

Two-Dimensional Color Palettes Generation

Zhang and Cox (1994) have tried several color palettes, which can be categorized as one of two types. In the first type, the colors were chosen to be sharply contrasting from point to point on the palette. This type is especially useful for a qualitative visual study of the wave pattern. For example, Figure 6 shows a palette that vividly enhanced the visual appearance of the wavy surface. This method, as illustrated in Figure 6, clearly brought out evidence of secondary waves normal to the propagation direction of long waves on the rear surface of long waves. In the second type, the colors were made to intermingle so that colors vary slowly and smoothly across the palette. This type of palette was found to be well adapted for analytical studies of the wave slopes because the continuous variation of colors can be interpreted in terms of a continuous variation of slopes. In order to understand why the continually varying color palette allows for better analytical measurements, it is important to realize that the light scattered from the water is not perfectly reflected or refracted, but is a diffused reflection or refraction. This will then smooth the boundaries between neighboring colors. If two neighboring colors on the palette are continuous, then the smoothed region is very close to either color and the error is minimized. If the colors are not continuous, smoothed regions between neighboring colors will average to a color which might be calibrated to a completely different slope. Naturally, this would result in an erroneous measurement.

Since the continually varying color palette allowed for superior analytical measurements, two palettes of this type were examined. For the first palette, the colors were

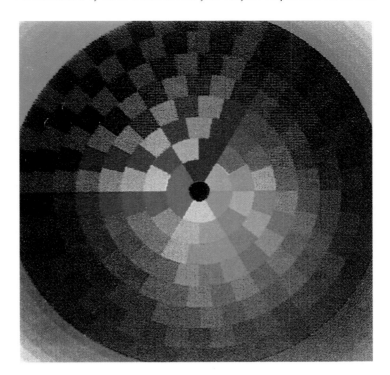

Figure 6 Color palette with noncontinuous colors.

presented in the RGB (red, green, blue) color space. Therefore, the colors for this palette were changed by varying the intensities of two of the three primary colors: red and green. The intensity of red was linearly varied in the vertical direction, the green was varied in the horizontal direction (Figure 7), while the intensity of blue was kept constant. In order to obtain accurate results from the RGB color palette, Zhang and Cox (1994) found it necessary to have diffused light of equal intensity at every position of the color palette. Unfortunately, this was quite a difficult task. Therefore, they followed the recommendations of Dabiri and Gharib (1991) to use hue and saturation rather than color for the slope coding for the second and more successful color palette. The colors for this palette were presented in the HSI (hue, saturation, intensity) color space (see Figure 8). Basically, hue is the essence of color expressed in the angular coordinate, while saturation represents the pureness of color. The hue and saturation together represent the quality of the color, while the intensity determines the quantity of the color. Dabiri and Gharib (1991), who used the HSI color space to carry out 2-D spatial thermometry for fluid flow experiments, present a discussion of the meaning of colorimetric terms, as well as a list of extensive references of published information. Thus, the colors for this palette were changed by varying the hue of colors in the azimuth of the slope and saturation of the colors according to the magnitude of the slopes. The intensity of the light was kept as constant as possible.

The amount of light intensity acquired with the camera depends on the refraction/reflection angle, depending on which mode is being used. For both modes, just beyond the maximum measurable slopes, the light intensity received by the camera is very small, regardless of the hue or saturation of the color being received. Therefore, if there are slopes just beyond the range of the system, the corresponding colors acquired will be black.

In order to obtain higher slope resolution, it is necessary to have more discrete colors. Since the hue and saturation bandwidth is limited, this implies that the color differences between hue and saturation steps must be smaller. The errors which contribute to the color differences between these steps are due to noise and to the spectral difference between the

Figure 7 Rectangular RGB color palettes.

light source and the camera (cross-coupling error). Thus, to ensure the highest level of confidence for the largest slope resolution, the color difference between two neighboring colors on the palette must be greater than the sum of the errors.

Figure 8 HSI color palette with hue varying in the azimuth and saturation varying in the radial directions. Note that unlike Figure 6, the colors vary in a continuous manner.

Data Analysis

There are several steps associated with processing the images in order to extract the slope information. First, the noise induced by the color camera's electronics and digitized into the images is basically "snow-like noise" in the spatial domain and "white noise" with a wide bandwidth in wavenumber domain. For this reason the digital images are first spatially filtered by a median filter to suppress the noise (Pratt, 1991). Second, when using a light source and a color camera together, there is a mismatch between the spectral sensitivity of the color camera and the light source. An extreme example of the damage that can be caused is that two beams of light generated by the color palette at two different positions, and therefore having different hue and saturation, can conceivably appear to have the same hue and saturation to the color camera. We call this effect "cross-coupling" of colors. Errors in slope due to cross-coupling can be apparent and a correction must be applied. For the RGB color space, Zhang and Cox (1994) assume a polynomial mapping between the light rays' directional cosine and the three-color component. Then through a series of calibrations, they are able to correct for the cross-coupling in the RGB color space. Similarly, for the RGB color space, it is very difficult to maintain constant intensity throughout the color palette. As such, color intensities are nonuniform in the image. Therefore, since the basic information in the RGB scheme is carried by the light intensity, errors will be introduced. The HSI scheme can do a better job of reducing these errors. First, if the intensity is not dim or overly bright, the hue is independent of the intensity. Second, the cross-coupling can be accounted for much more easily in the HSI color space as will be shown below.

Zhang and Cox (1994) perform the transformation from color to slope components in two steps. First, the color is mapped to the angle of the light ray underwater. In this step, cross-coupling corrections are taken into consideration. Second, ray angles are mapped to the slope components of the water surface based on geometrical optics.

The surface gradient can be written as

$$\nabla f = \|\nabla f\| \cdot e^{i\phi} = (f_x, f_y) \qquad (9)$$

where f represents surface height, and $\|\nabla f\|$ and ϕ are the magnitude and direction of the surface gradient. Corresponding to each surface gradient, $(\|\nabla f\|, \phi)$, there are rays in a parallel beam directed from the underwater lens toward the water surface whose directional cosines are $\alpha_x = A\cos(\Phi)$, $\alpha_y = A\sin(\Phi)$. The tangent of the vertical, A, is related to $\|\nabla f\|$ by Snell's law. Therefore, let $\theta_0 = \tan^{-1}\|\nabla f\|$ and $\theta_1 = \tan^{-1}\|A\|$, and

$$\theta_0 = \theta_1 - \sin^{-1}[\sin(\theta_1)/n] \qquad (10)$$

For the color screen shown in Figure 8, the origin of rays on the screen is such that Φ is a function only of hue (H) and A is a function of saturation (S). The mapping is then assumed to be

$$\Phi = \hat{H} = \sum_{i=0}^{6} d_i H^i \qquad (11)$$

and

$$A = \hat{S} = \sum_{m=0}^{3} a_m^0 S^m \left\{ 1 + \left[\sum_{n=1}^{3} \left(\cos(n\Phi) b_n + \sin(n\Phi) C_n \right) \right] \right\} \qquad (12)$$

where the coefficients a, b, c, and d are determined from the calibration data. The term in curly braces is used to correct for the cross-coupling. Corrections for cross-coupling of the saturation within the hue data (Equation 11) were not done, because calibrations indicated this is not important.

The second step was to map from the underwater ray angle to the slope of the water surface using geometric optical ray tracing. If the vertical angle of the underwater ray is θ_0 and the slope angle of the water surface is θ_1, the relation between the two is given in Equation 4.3. Since there is no analytical solution for the inverse of this relation, a numerical solution was used by Zhang and Cox (1994).

Derivation of the Free Surface Elevation

Once the free surface slopes are determined, the free surface elevations can be found through direct integration.

$$f(x,y) = \int \nabla f(\vec{x}) \cdot d\vec{x} \tag{13}$$

where $f(x,y)$ is the surface elevation and ∇f is the free surface slope. Ideally, there are many different integral paths that lead to the same result. At this stage, however, we integrate through the use of the Fourier transform.

If the surface elevation is represented by $f(x,y)$, its two orthogonal slope components are represented by $f_x(x,y)$ and $f_y(x,y)$; and their Fourier transforms are $F(k_x,k_y)$, $F_x(k_x,k_y)$, and $F_y(k_x,k_y)$, respectively, where k_x and k_y are the two orthogonal wave number components. Then the following relations hold (Zhang and Cox, 1994):

$$f(x,y) + C \Leftrightarrow F(k_x, k_y) = \frac{F_x(k_x, k_y) \cdot (-ik_x) + F_y(k_x, k_y) \cdot (-ik_y)}{k^2}, \ k \neq 0 \tag{14}$$

where the double-headed arrow represents a Fourier series transformation. The final expression produces the elevation of the water surface from the transforms of the slope components. Unfortunately, the constant of integration, C, cannot be determined from slope measurements.

Zhang and Cox (1994) clearly point out two points. First, one must be careful to avoid 2-D image aliasing. This can be done by making sure that the bandwidth of data is less than half the sampling frequency. Second, the periodic extension implied by the Fourier series transformation creates artificial discontinuities at the edges of the field of view. In order to reduce this effect, one can multiply the field by a window function so that the slopes fade to zero smoothly toward the field of view's boundary. A second method is to extend the data artificially to a larger area such that the boundary values smoothly approach a constant. The periodic extension is then carried out for this larger area. This approach will then avoid data discontinuities and will preserve correct data analysis up to the edges of the image, at the expense of extra time for calculations in the artificially extended regions. Finally, in order to have smooth boundaries for both the water elevation and the slope, the constant of integration should be such that it cancels out the mean slope.

Applications

Aside from providing quantitative information regarding the free surface slopes and elevations, the present technique produces aesthetically beautiful flow visualization pictures. Figure 9 shows the free surface of a shear layer just after a surface-piercing splitter plate, where Re = 2000 and the flow is from top to bottom. These images were acquired in the

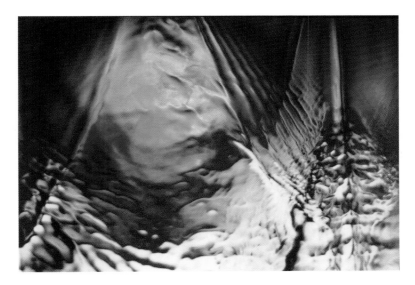

Figure 9 Free surface of a shear layer just after the surface-piercing splitter plate. The flow is from top to bottom and Re = 2000.

refractive mode since the slopes of the waves are quite large. Figure 10 shows an air jet blowing over the free surface of a water tank. This image was also acquired in the refractive mode since the slopes of the free surface waves are large. The jet nozzle can be seen at the left of the image and the jet flow is from left to right. Note that the pattern of the waves is in an elliptical path surrounding the jet flow over the free surface. Zhang (1994; 1995) has been interested in studying capillary gravity and capillary waves on surface oceans. Since

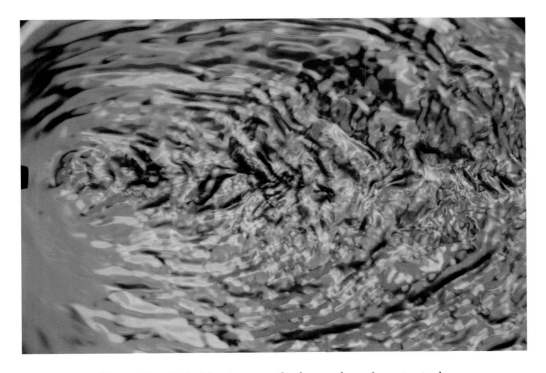

Figure 10 Air jet blowing over the free surface of a water tank.

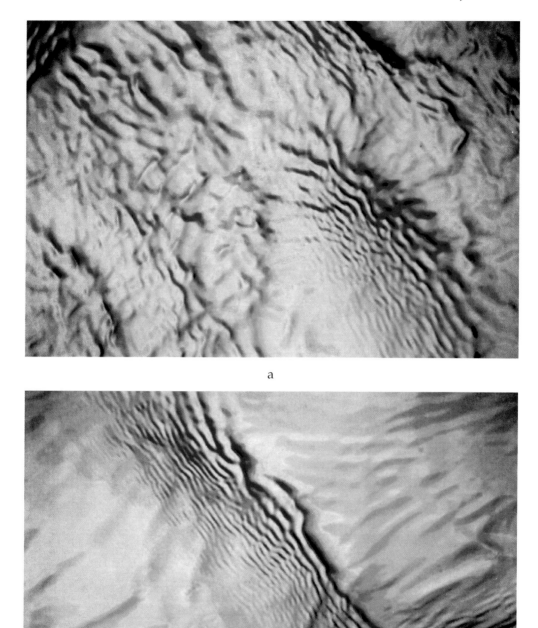

Figure 11 (a and b) Images of wind-generated capillary gravity and capillary waves.

these waves generally have large slopes, the refractive mode was used to acquire images of the free surface. Two of these images are shown in Figures 11a and 11b.

Figure 12 Interaction of a vortex tube connected with a free surface and surface waves.

Unfortunately, since the free surface color mapping technique is relatively new, few applications have made use of this technique. However, there are a few applications which demonstrate the technique quite effectively. In an attempt to understand the interactions of free surface waves with near-surface vorticity, Zhang et al. (1994) generated a vortex tube connected to the free surface while passing waves over the vortex. Figure 12 shows an acquired image from this setup. It is important to realize that this image is the raw unprocessed image. Multiple lens arrays (Figure 3) were used to image an area of 22×14 cm^2. The color palette used is shown in Figure 13. Since the dimple caused by the vortex on the free surface contains all slopes, the colors acquired from the dimple will have all colors. Thus, the dimple on the color image can be identified from the "color pie" seen in the lower center of the picture. The wave motion is from the lower right to the upper left of the picture. The wave fronts can be identified by the lines of constant color. Figure 14 shows the calculated surface elevations. The deformation of the free surface in the form of a dimple is clearly seen. Also, the waves are seen moving with fronts perpendicular to the vortex. Note that this figure clearly shows that, as the waves pass over the vortex, discontinuities are created between the sequential wave fronts.

In order to understand the interaction of turbulence with a free surface, Dabiri et al. (1994) measured the surface elevations of grid-generated turbulence interacting with a free surface. The reflective mode was used since the slopes ranged from -4 to $4°$ and was therefore small in range. Figures 15 and 16 show the raw images acquired at 0.1 s apart using the color palette shown in Figure 13. As in the previous example, multiple lens arrays were used to image a 22×15-cm^2 portion of the free surface. Figures 17 and 18 show the surface elevations corresponding to Figures 15 and 16 calculated from the measured slopes. Note that the elevation range is small and is under 1 mm for both cases. The time sequence of these images shows some interesting behavior of the vorticity

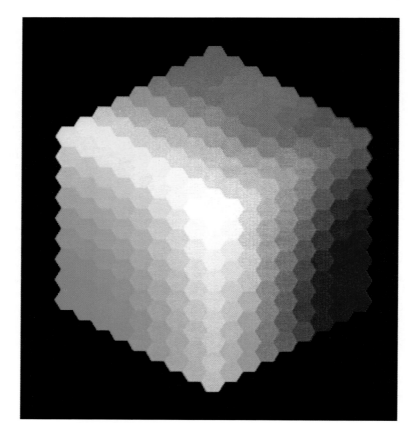

Figure 13 The color palette used to color code the free surface slopes of Figures 12, 15, and 16.

connected to the free surface. Vortices 1 and 2 remain stationary while its relative depression has changed. Vortices 3 and 4 in Figure 17 have merged into one vortex labeled as vortex 3 in Figure 18. Figures 19 and 20 show the 3-D surface plot of a portion of the image squared out in Figures 17 and 18, where a vortex tube has connected with the free surface, thus causing a dimple on the free surface. Figure 21 shows the spectra of the free surface waves with 1.5-cm wavelengths through time. The solid line shows the viscous wave theory. Note that the experimental results follow the theory until 4 s, beyond which the experimental spectra remain flat.

Conclusion

The free surface color mapping technique described in this paper is a powerful tool of deriving the free surface slopes and elevations in real time. The technique is based on generating a system of color beams, where each set of color rays is parallel to themselves, yet at different angles with respect to rays of other colors. This system of beams is then used to illuminate the free surface to be measured. There are two modes which can be used. If the ranges of slopes to be measured are less than ~10°, then the reflective mode is recommended. If the range of slopes to be measured is at most ~50°, then the refractive mode is recommended. For data acquisition, refracted rays from the free surface are captured for the refractive mode. Earlier versions of this technique acquired the data onto photographic film. However, the present technique acquires the data through a three-chip color camera onto three separate laser disk recorders to maintain

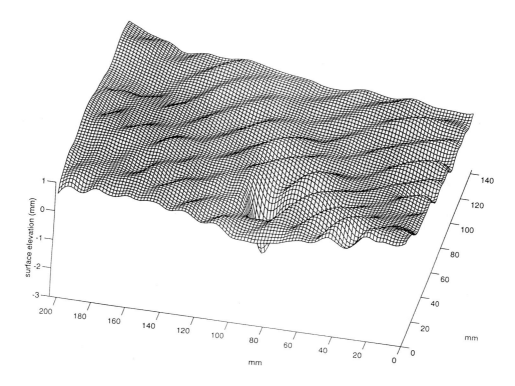

Figure 14 The surface elevation calculated from the slope obtained from Figure 12.

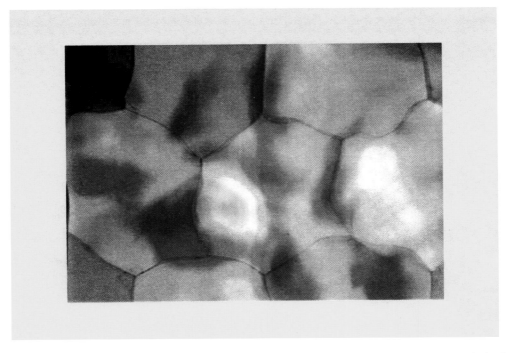

Figure 15 Image of a free surface showing the interaction of grid-generated turbulence with the free surface.

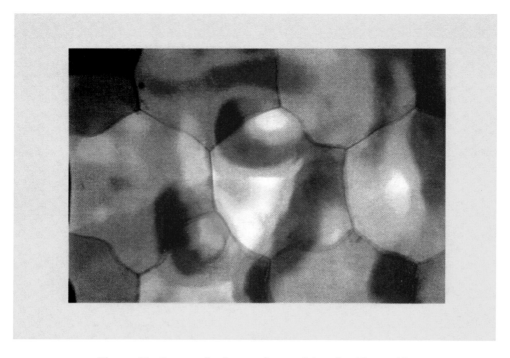

Figure 16 Image of a free surface at 0.1 s after Figure 15.

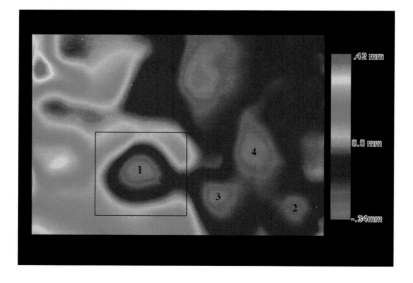

Figure 17 Surface elevations of the free surface corresponding to Figure 15.

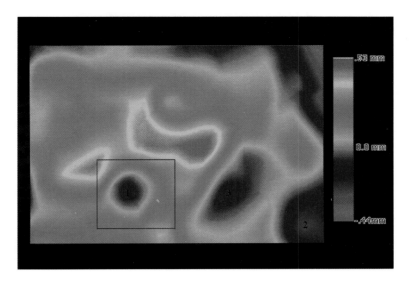

Figure 18 Surface elevations of the free surface corresponding to Figure 16. Note the change of the numbered vortices.

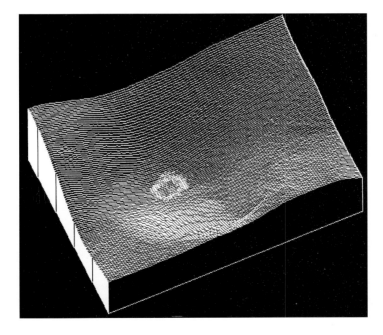

Figure 19 The 3-D plot of surface-connected vortex shown in the square in Figure 17.

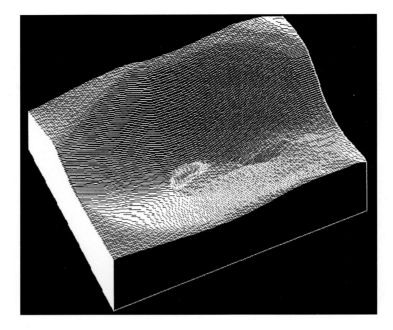

Figure 20 The 3-D plot of surface-connected vortex shown in the square in Figure 18.

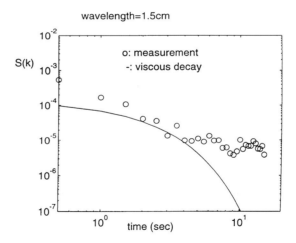

Figure 21 Spectra of the free surface for 1.5-cm wavelengths. The solid line indicates results obtained from viscous wave theory.

high bandwidth data acquisition. The fundamentals of the algorithms used to calibrate and extract slopes and integrate the slopes into surface elevations have also been explained. Last, applications of the technique are shown for both the reflective and refractive modes.

Acknowledgments

The authors would like to thank Ed Rood and the Office of Naval Research Fluid Dynamics Program for supporting this work (ONR-URI, Grant No. N00014-92-J-1610).

References

Banner M.L. et al.; Wavenumber spectra of short gravity waves, *J. Fluid Mech.*, 198, 321, 1989.
Barber N.F.; A diffraction analysis of a photograph of the sea, *Nature*, 164, 485, 1949.
Cox C.S.; Measurement of slopes of high-frequency wind waves, *J. Mar. Res.*, 16, 199, 1958.
Cox C.S.; Munk W.; Statistics of the sea surface derived from sun glitter, *J. Mar. Res.*, 13, 198, 1954.
Dabiri D.; Gharib M.; Digital particle image thermometry: the method and implementation, *Exp. Fluids*, 11, 77, 1991.
Dabiri D.; Zhang X.; Gharib M.; A real-time free surface elevation mapping technique, in *7th Int. Symp. on Applications of Laser Techniques to Fluid Mechanics*, Lisbon, Portugal, July 11–14, 1994.
Ebuchi N.; Kawamura H.; Toba Y.; Fine structure of laboratory wind-wave surfaces studies using an optical method, *Boundary Layer Meteol.*, 39, 133, 1987.
Holthuijsen L.H.; Stereophotography of ocean waves, *Appl. Ocean Res.*, 5, 204, 1983.
Jahne B.; Waas S.; Optical measuring technique for small-scale water surface waves, in *Advanced Optical Instrumentation for Remote Sensing of the Earth's Surface, SPIE Conf. Proc.*, Washington, D.C., 1129, 122, 1989.
Jahne B.; Riemer Klaus S.; Two-dimensional wave number spectra of small-scale water surface waves, *J. Geophys. Res.*, 95, 11531, 1990.
Jahne B.; Klinke J.; Waas S.; Imaging of short ocean wind-waves — a critical review, *J. Opt. Soc. A*, 11(8), 2197, 1994.
Kasevich R.S.; Directional wave spectra from daylight scattering, *J. Geophys. Res.*, 80, 4535, 1975.
Keller W.; Gotwols B.L.; Two-dimensional optical measurement of wave slope, *Appl. Opt.*, 22, 3476, 1983.
Kohlshutter E.; Die Forschungareise S. M. S. Planet Band 2, Stereophotogrammetrische Aufnahmen, *Ann. Hydrogr.*, 34, 219, 1906.
Kraichnan R.H.; Inertial-range transfer in two-dimensional and three dimensional turbulence, *J. Fluid Mech.*, 47, 525, 1971.
Laas W.; Photographische Messung der Meereswellen, *Vereins Deutsch. Ingenieure*, 49, 1889, 1937, 1976, 1905.
Longuet-Higgins M.S.; Reflection and refraction at a random moving surface. I. Pattern and paths of specular points, *J. Opt. Soc. Am.*, 50, 838, 1960.
Lubard S.C. et al.; Optical image and laser slope meter intercomparisons of high-frequency waves, *J. Geophys. Res.*, 85, 4996, 1980.
Lugt H.J.; Local flow properties at a viscous free surface, *Phys. Fluids*, 30, 3647, 1987.
Pratt K.P.; *Digital Image Processing*, 2nd ed., John Wiley/Interscience, New York, 1991.
Schooley A.H.; A simple optical method for measuring the statistical distribution of water surface slopes, *J. Opt. Soc. Am.*, 44, 37, 1954.
Stilwell D.; Directional energy spectra of the sea from photographs, *J. Geophys. Res.*, 74, 1974, 1969.
Tober G.; Anderson R.C.; Shemdin O.H.; Laser instrument for detecting water ripple slopes, *Appl. Opt.*, 12(4), 788, 1973.
Wu J. et al.; A multiple purpose optical instrument for studies of short water waves, *Rev. Sci. Instrum.*, 40, 1209, 1969.
Zhang X.; Capillary-gravity and capillary waves generated in a wind wave tank: observations and theories, *J. Fluid Mech.*, 289, 51, 1995.
Zhang X.; Cox C. S.; Measuring the two-dimensional structure of a wavy water surface optically: a surface gradient detector, *Exp. Fluids*, 17, 225, 1994.
Zhang X.; Dabiri D.; Gharib M.; A novel technique for free-surface elevation mapping, *Phys. Fluids*, 6, S11, September 1994.

chapter two

Dynamic and Multiple Laser Light Sheets: An Experimental Approach of Unstationary and Three-Dimensional Flows

J.P. Prenel, R. Porcar,[1] A. Texier,[2] and A. Strzelecki[3]

[1]IGE, Université de Franche-Comté, Belfort, France
[2]LMF, Université de Poitiers, Poitiers, France
[3]Cert-Dermes Onera, Toulouse, France

> **Abstract**—*We present the recent improvements of the laser tomography techniques. After a theoretical comparison between static and dynamic illumination modes, we describe various methods making use of laser light sheets.*
>
> - *Dynamic sheets generated by means of one-dimensional laser beam sweeps allow the two-dimensional analysis of instabilities. An example of quantitative analysis of shock waves oscillations is presented.*
> - *Movable plane sheets generated by the controlled shift of a static sheet lead to sequential or volumic visualizations of three-dimensional flows. Examples of three-dimensional vortices around a Delta wing or in a curved duct are shown.*
> - *Simultaneous multiple plane sheets can be generated by means of a selective wavelength coding tomography. A polychromatic set of three parallel sheets gives an approach of the three-dimensional velocity field of an unstationary water flow.*

Introduction — Basic Laser Light Sheets

The study of flows by tomographic means, born in the 1970s (Schneiderman and Sutton, 1970; Philbert and Boutier, 1972; Philbert et al., 1979; Porcar, et al., 1975), is now well known in all fields of fluid mechanics. Most of the time a sole plane is studied with a single laser light sheet generated by the spreading of a focalized beam (Merzkirch, 1987). Optical systems allowing one to realize plane laser sheets have been described in various articles (Porcar et al., 1983; Diemunsch and Prenel, 1987). We are reminded here of the main scheme: a standard optical device (most often a single lens, sometimes an adjustable telescope) allows the focusing function. A cylindrical optical device (single lens or telescope) allows the spreading function, according to one direction, to make a focal line from the focal point (Figure 1). We can note that the usual terminology (plane sheet–sheet thickness) is, in fact, ill-advised; the real sheets are complex light distributions and their

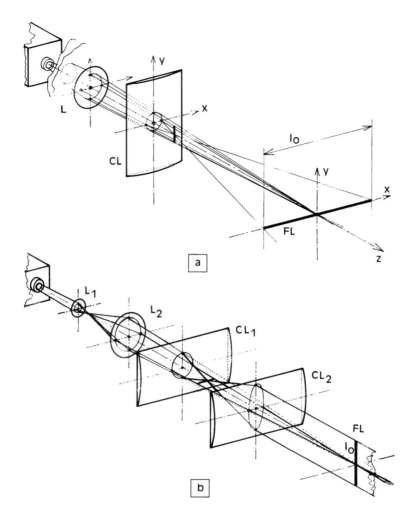

Figure 1 Basic laser sheet generators: focusing–spreading setup. (a) Classical divergent light sheet. L = spherical lens, CL = cylindrical lens, FL = focal line. (b) Light sheet with constant width. $L_1\ L_2$ = spherical telescope, $CL_1\ CL_2$ = cylindrical telescope.

characteristic parameters can easily be calculated only if the illuminating laser operates on a TEM_{00} mode (Figure 2). In particular, the minimal thickness E_0 on the optical axis is equal to the beam waist diameter $2w_0$ at e^{-2}, which is given by the Kogelnik gaussian beam theory (Kogelnik, 1966):

$$\frac{1}{w_0^2} = \frac{1}{f^2}\left(\frac{\delta_c^2}{w_c^2} + \frac{\pi^2}{\lambda^2} w_c^2\right) \tag{1}$$

where w_c is the laser cavity waist radius at e^{-2}. The focusing half-angle α is a function of the wavelength λ and of the waist radius w_0.

$$tg\ \alpha \approx \frac{\lambda}{\pi w_0} \tag{2}$$

Chapter two: Dynamic and multiple laser light sheets

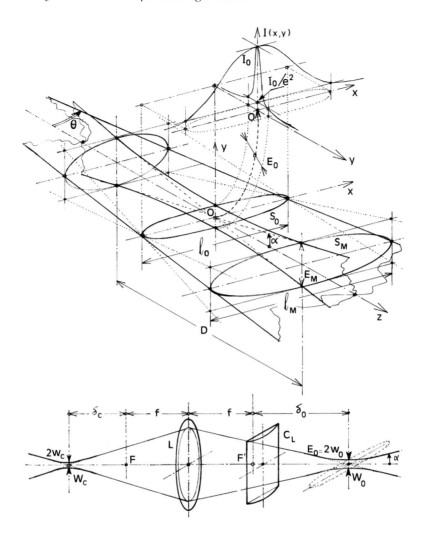

Figure 2 Light distribution in a basic divergent laser sheet (gaussian beam). W_c = cavity waist (radius w_c), W_0 = lens waist (radius w_0), f = lens focal distance (focusing), θ = divergence half-angle, α = focusing half-angle, l_0 = sheet width, E_0 = sheet thickness, D = field depth, I = light power density.

Usually, the divergence half-angle θ and thus the sheet width l_0 (always at e^{-2}) are deduced from the elementary first-order optics theory. They are principally a function of the focal distance of the spreading lens CL (Figure 1a).

The practical depth of field D is chosen by the operator. The D value is deduced from the maximal "acceptable" sheet thickness E_M on the optical axis, according to the gaussian beam theory:

$$w(z) = w_0 \sqrt{1 + \frac{z^2}{z_R^2}} \qquad (3)$$

where $z_R = \pi w_0^2 / \lambda$ is the Rayleigh distance.

If we wish a particular E_M value, the field depth can be written as follows:

$$D = 2z_R \sqrt{\frac{E_M^2}{E_0^2} - 1} \qquad (4)$$

where $E_M = 2w(D/2)$
$E_0 = 2w_0$

If $E_M = E_0 \sqrt{2}$, D becomes equal to $2z_R$. It is important to notice that a large depth of field and a very small E_0 value are incompatible because of the properties of a gaussian beam waist.

For example, if we retain the above hypothesis ($E_M = E_0 \sqrt{2}$) we obtain typical D values:

$w_0 = E_0/2$ (µm)	$D = 2z_R$ (mm)
10	1.22
50	30
100	122
300	1100

A usual criterion to determine the "acceptable" E_M value is the choice of the minimal light power density at the $D/2$ distance from the central zone. For an incident gaussian beam, the power density is given by

$$I(w) = I_0 \exp\left[-\frac{2w^2}{w_0^2}\right] \qquad (5)$$

where w_0 is the e^{-2} characteristic beam radius. The total beam power P_t being known, we can calculate I_0 from

$$P_t = \int_0^\infty I(w) \cdot 2\pi w \, dw = I_0 \frac{\pi w_0^2}{2} = I_0 \frac{\sigma_0}{2} \qquad (6)$$

where πw_0^2 is the characteristic beam area at e^{-2}.

The light power present in this area, σ_0, can be written as follows:

$$P_0 = \int_0^{w_0} I(w) \cdot 2\pi w \, dw = I_0 \frac{\sigma_0}{2}\left(1 - e^{-2}\right) \approx 0.865 \, P_t \qquad (7)$$

When the beam is spread by an ideal cylindrical lens we can admit that the power P_0 appears in the elliptical section S_0 (Figure 2) (the losses in the refracting interfaces can be deduced). So we can approach the mean power density in the central zone of the sheet by

$$\overline{I_0} \approx \frac{P_0}{S_0} = \frac{P_t(1-e^{-2})}{\pi \, E_0/2 \, l_0/2} \qquad (8)$$

or determine this mean value in the S_M section ($z = D/2$).

$$\bar{I} \approx \frac{P_0}{S} = \frac{P_t(1-e^{-2})}{\pi \, E_m/2 \, l_M/2} \tag{9}$$

When a simple qualitative visualization is wished, the decrease of the power density is not very embarrassing and the lighting criterion not very strict. If a quantitative aspect and an image processing are wished, the power density decrease must be limited and the criterion becomes more rigorous: for example, $\bar{I}/I_0 > 0.8$ can be a good compromise.

Plane Dynamic Sheets by Means of Beam Sweeps

Increasing of the Transverse Lighting Homogeneity

If the spreading function is replaced by a sweeping function, the laser sheet becomes dynamic. The cylindrical optical setup is suppressed and the focalized beam is reflected by a movable mirror or prism, mounted on a rotating device (Ayrault et al., 1980) or on a ferromagnetic galvanometer, including a position transducer to assure servo-controlled operations. The driving signal of this optical scanner is usually sinusoidal, or triangular if a uniform sweeping speed is desired (Figure 3a). The focal line becomes circular, most often approximated by a rectilinear segment, when the focusing distance is large. If necessary, the rotation movement of the beam can be transformed into a parallel shift by means of a collimating lens L3 (Figure 3a). In this configuration, the sheet width l_0 is naturally limited by the lens diameter, but it becomes easy to illuminate an internal flow in a cylindrical duct without meeting the transparent walls.

A typical example is presented in Figure 4a. A stationary shock wave system is generated in a supersonic ejector (Mach number about 2). In spite of the presence of an air dryer, the large expansion ratio in the ejector gives a water condensation spray. The concentration of submicronic droplets is very low and the spray remains invisible by direct observation; but the local concentration, upstream of the shock surfaces, is sufficient to be revealed by the scattered light because of the high power density of the sweeping laser beam.

The main advantages of this dynamic technique are the good flexibility of the sheet width adjustment and principally the more homogeneous transverse light distribution. As a matter of fact, if the laser beam shift along the x axis (Figure 3a) is periodic between $x = -a$ at $t = 0$ and $x = +a$ at $t = T/2$, the instantaneous power density becomes

$$I(x,y,t) = I_0 \exp\left(\frac{-2y^2}{w_0^2}\right) \exp\left[-2\frac{(x-r)^2}{w_0^2}\right] \tag{10}$$

where $r(t)$ is related to the sweep velocity v(t).

$$r(t) = \int_0^t v(t)dt$$

For a triangular driving, $r(t) = a\,(4ft - 1)$, where $f = 1/T$ is the sweep frequency and a the sweep amplitude. For a sinusoidal one

$$r(t) = -a \cos(2\pi ft)$$

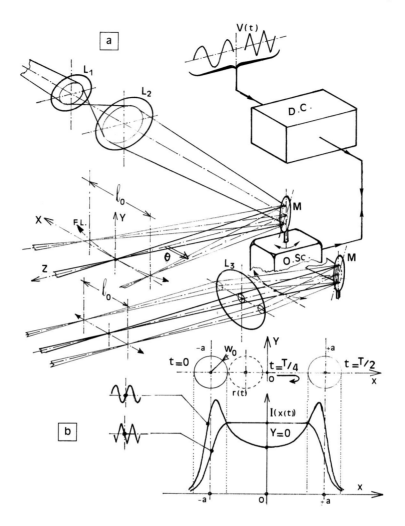

Figure 3 Plane dynamic laser sheets. (a) Optical setup. $L_1 L_2$ = focusing function, 0.Sc. = optical scanner: sweeping function, M = oscillating mirror, D.C. = driving control, V(t) = driving signal, L_3 = collimating lens. (b) Typical calculated power density distribution. a = sweep amplitude ($\ell_0 \approx 2a$).

The elementary energy $d^3\varepsilon(xy)$ received by an area $dx\,dy$ around $M(x,y)$ during dt is written as follows:

$$d^3\varepsilon(x,y) = I(x,y,t) \cdot dx\,dy\,dt \tag{11}$$

The local power density is deduced by a time integration during a half-period T/2.

$$I(x,y) = 2I_0 f \cdot \exp\left(-\frac{2y^2}{w_0^2}\right)\int_0^{T/2} \exp\left[-2\frac{(x-r)^2}{w_0^2}\right]dt \tag{12}$$

For a triangular driving, the simple form of the gaussian integrand gives a formal expression containing the error function

Chapter two: Dynamic and multiple laser light sheets

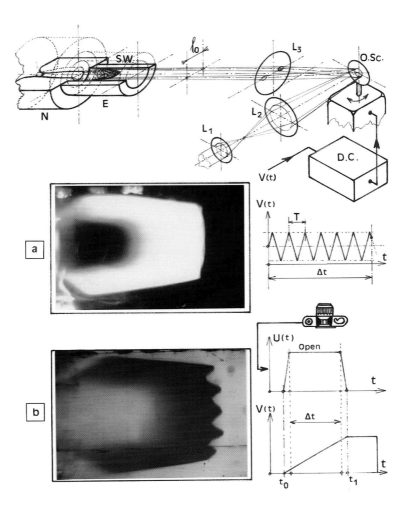

Figure 4 Dynamic plane laser sheet: supersonic internal flow visualization. N = nozzle, E = supersonic ejector, S_w = shock wave system, L_3 = collimating lens ϕ = 10 cm, O.Sc. = optical scanner, D.C. = driving control, V(t) = driving signal, U(t) = triggering signal, Δt = exposure time. (a) Stationary supersonic mode: periodic illumination. Argon laser P = 3 W, Δt = 1/50 s, f = 1/T = 400 Hz, 400 ISO emulsion, ℓ_0 = 8 cm. (b) Unstationary supersonic mode: single sweep illumination. Δt = 0.2 s, 1600 ISO emulsion.

$$I(x,y) = \sqrt{\frac{\pi}{2}} \frac{P_t}{2\pi w_0 a} \exp\left(-\frac{2y^2}{w_0^2}\right) \left[erf\left(\sqrt{2} \cdot \frac{a-x}{w_0}\right) + erf\left(\sqrt{2} \cdot \frac{a+x}{w_0}\right) \right] \qquad (13)$$

An example of spatial profile $I(x,0)$ is presented in Figure 3b; it shows a flat power density distribution between $-a$ and $+a$.

For a sinusoidal driving,

$$I(x,y) = 4 f P_t \exp\left(-\frac{2y^2}{w_0^2}\right) \int_0^{T/2} \exp\left[-2\frac{[x + a\cos(2\pi ft)]^2}{w_0^2}\right] dt \qquad (14)$$

A numerical integration allows to plot the power density profile; for example, $I(x,0)$ shows a flat light distribution near the central zone and two lateral peaks due to the fact that the mean presence time is longer near the boundaries.

The theoretical advantage of the triangular sweep is evident; but we must notice that the real movement of the oscillating mirror is rarely strictly triangular except for very low sweep frequencies.

When this frequency increases up to a few tens of hertz, little lateral peaks appear near the border of the light profile. In spite of this small shortcoming, the conclusion is obvious: in any case where the stroboscopic effects due to the beam movement are without importance (stationary or slowly variable phenomena — Figure 4a), the choice of a dynamic sheet is preferable because of its homogeneous light distribution.

However, if the driving frequency of the optical scanner is close to a characteristic frequency of the studied phenomenon, the image interpretation can be altered. But this disadvantage can become an advantage if we suitably exploit these stroboscopic effects.

Optical Oscilloscope: Visualization of Supersonic Instabilities

Taking one's inspiration from the classical oscilloscope we can use a single, constant velocity beam sweep (Diemunsch et al., 1985). If this beam movement is orthogonal with a shift of the shock structure, this one can be quantitatively analyzed. The traveling time of the laser pencil is easily adjustable to the frequency of the phenomenon to be investigated. The triggering is assured by the recording device. For example, a classical photographic camera unlocks the sweeping signal by means of its flashlight switch (Figure 4b). The fixing of the laser pencil in its furthest position during the required time of shutter closing avoids superimposition of images caused by the fast beam return. This technique is easily adapted to the recent one-shot CCD cameras.

An example is presented in Figure 4b. An unstationary supersonic mode is generated by decreasing the generative pressure of the nozzle; the Mach disk of the shock structure becomes oscillating. The oscillation function is clearly visible on the picture. The measurement accuracy of the vibration amplitude is approximately 5% for a measured frequency of about 20 Hz.

Three-Dimensional (3-D) Approach: Two-Dimensional (2-D) Beam Sweeps and Movable Plane Sheets

Two-Dimensional Beam Sweeps

The combination of two orthogonal optical scanners allows to generate 3-D dynamic light sheets. The shape of the sheet cross-section depends on the choice of the electrical signals driving each scanner, the position and velocity of which are controlled by a closed loop. The illumination mode can be selected as single or multiple paths (Figure 5).

An example of this application is shown in Figure 5 (Prenel et al., 1986). The shock structure presented in Figure 4a becomes nonaxisymmetrical because of a parietal air injection (lower zone). Two quasi-simultaneous orthogonal sections of the supersonic flow are visualized by means of a cross-shaped dynamic sheet, the sweeping frequency being about 400 Hz; the illumination mode is of the multiple path type.

As for 1-D beam sweeps (see previous section), the laser pencil movements can induce stroboscopic effects and lead to erroneous interpretations. If this problem is too embarrassing, it is preferable to return to the permanent lighting of a static sheet. In this case the third dimension can be investigated only by a shift of the sheet.

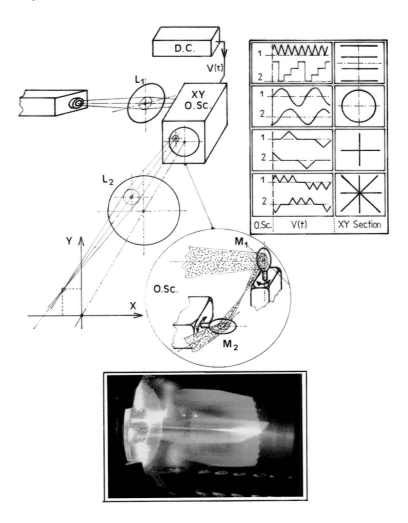

Figure 5 2-D beam sweeps. Visualization of two orthogonal sections of a nonaxisymmetrical shock structure. L_1L_2 = focusing function, M_1M_2 = oscillating mirrors, XY 0.Sc. = two-way optical scanner, D.C. = driving control. Exposure time, 10^{-2} s; sweep frequency, 400 Hz; Argon laser P = 3 W; 800 ISO emulsion.

Movable Plane Sheets

Another approach of 3-D visualizations is the combination of the permanent lighting of a spreadsheet, which facilitates the image recording and interpretation with a fast movement of this sheet, generated by an optical scanner (Prenel et al., 1988; Prenel et al., 1989; Yoda and Hesselink, 1990).

This scanner can be set near the front focal plane of the second telescope lens L_2, to convert the beam rotation into a translation movement (h) (Figure 6). This configuration induces a shift δq of the focal line during the mirror rotation. According to the first-order optics theory, this shift is written as follows:

$$\delta q = -f_2^2 \frac{1-\cos 2\beta}{(p-f_2)\cos 2\beta} \tag{15}$$

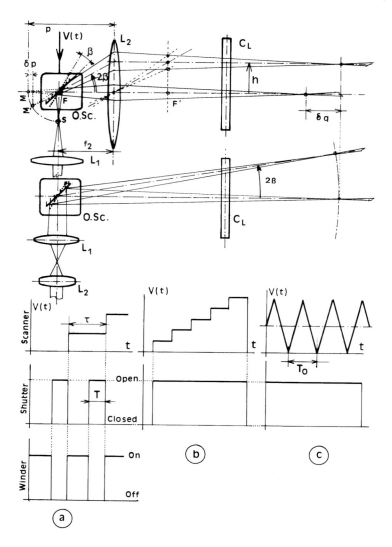

Figure 6 Movable plane laser sheets. (a) Synchronized sequential illumination; (b) superimposed sequential illumination; (c) volumic periodic illumination.

where

β = the rotation angle of the scanner
f_2 = the L_2 focal distance
p = the distance between L_2 and the virtual punctual source Σ

This shift remains weak for usual configurations ($\beta \approx 15°$, $\delta q \approx 50$ to 100 mm) and is more often compensated by the high field depth. In any case, the h value is always limited by the diameter of the lens L_2.

A second solution consists of setting the scanner behind the telescope and lighting the cylindrical spreading lens CL with a variable angle 2β. Then the sheet movement becomes a rotation, but wide shifts are possible if necessary, particularly for wind tunnel experiments. A computer drives the scanner; the motion mode determines the type of visualization completely (Figure 6).

Synchronized Sequential Illumination (Figure 6a): The scanner is driven by a step function and a photographic or video recording can be made during each intermittent stop of the sheet. The choice of the recording system determines the possible rate of taking tomographic "snapshots". If a standard photocamera is used, its winder speed (a few images per second) imposes the frequency of recording and the film shift is triggered by the flash connection. If a standard CCD camera is used, the time gap between two recordings is about 20 ms and the triggering operation is carried out by the video composite signal. The focusing function during the recording is ensured by the auto focus system, but the experimenter must cautiously choose the depth of field of the objective.

Superimposed Sequential Illumination (Figure 6b): If a photographic camera is used, the winder motor can be disconnected and the shutter can remain open while the sheet performs N successive translation movements. The pause time τ must be adapted to the sensitivity of the photographic emulsion. If the sheet movement time is negligible in comparison with τ, the film presents the superimposed images of N cross sections of the flow, allowing the reconstruction of its spatial evolution.

Periodic Volumic Illumination (Figure 6c): If the scanner is driven by an alternative signal, the sheet periodically oscillates and the visualization becomes continuous in the sweeped volume. The image contrast strictly depends on the mean power density of the light sheet, then on its width and its thickness. The driving signal can be sinusoidal or triangular as for plane beam sweeps (see last section: "Plane Dynamic Sheets by Means of Beam Sweeps"). A sinusoidal signal induces a nonlinear motion so the electromechanical scanners can tolerate higher frequencies (up to 1 kHz). The triangular signal warrants a linear sweep, then a good homogeneity of the illumination (like in Figure 3b), and constitutes a good compromise for frequencies (about a few hundred hertz) as for sweep amplitudes.

We present two examples of application:

1. Air flow around a Delta wing at low Reynolds numbers in a wind tunnel (Figures 7 and 8).

The Delta model is 10 cm wide with a variable incidence angle. The air seeding is ensured by means of a micronic oil aerosol, the injection velocity of which is strictly adapted to the air stream. The seeding mode can be punctual (Figure 8a and 8b) or matrical (5 × 5 injections — Figure 8c). Two sweeping modes are shown here:

- A superimposed sequential illumination (Figure 8a) — five successive planes are recorded: the pause time of the sheet is $\tau = 0.25$ s and the distance gap between two planes is about 1 cm.
- A periodic volumic illumination — the sweeping frequency is fixed at 100 Hz and the sweeping amplitude at about 15 cm.

For these different visualizations, the sheet thickness is cautiously optimized about 200 μm and the depth of field is chosen to be strictly adapted to the wind tunnel width.

2. Water flow downstream of a bend (Cabasset et al., 1995) (Figures 7 and 9).

The test rig (Figure 7) is a hydrodynamic bench with a maximal flow rate of 140 m^3/h. A closed tank (25 diameters upstream of the test section) generates a Poiseuille flow with

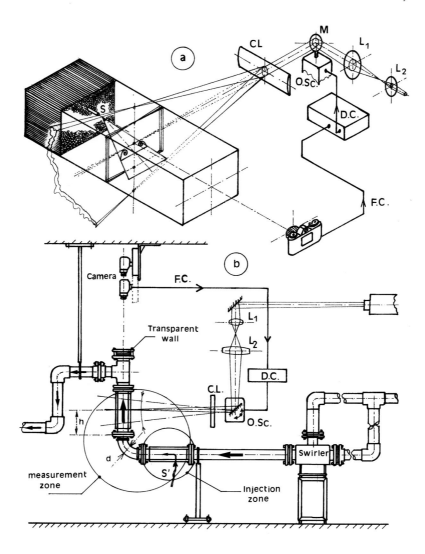

Figure 7 Synchronized sequential tomographic illumination. (a) Air flow around a delta wing, S oil droplets seeding; (b) water flow downstream of a bend, S' dye seeding.

a mean velocity between 0.01 and 0.02 m/s in a straight pipe before the model. This model is a perspex bend: radius of curvature $R = 150$ mm and internal pipe radius $R_0 = 50$ mm. The seeding is ensured by a fluorescent dye injection, 5 diameters upstream of the model.

The unstable laminar flow in the bend, consequence of the imbalance between the centrifugal forces and the radial pressure gradient, is characterized by the Dean number:

$$K = R_e \sqrt{\frac{R_0}{R}} \tag{16}$$

where R_e is the Reynolds number related to the pipe diameter $d = 2R_0$.

In the bend, the flow presents several pairs of contrarotating vortices: Dean, Görtler, and internal vortices. The Dean vortices are localized inside of the bend, the Görtler ones

Chapter two: Dynamic and multiple laser light sheets   35

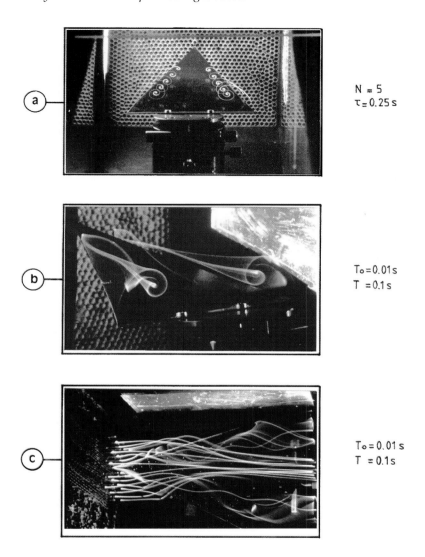

Figure 8 3-D visualizations at low Reynolds numbers in a wind tunnel around a delta wing. 480 < Re < 1920. (a) Superimposed sequential illumination, punctual seeding; (b) periodic volumic illumination, punctual seeding; (c) periodic volumic illumination, matrical seeding 5 × 5. Argon laser P = 2 W; 1600 ISO emulsion; micronic oil droplets seeding.

outside. Different authors are not in agreement on the number of vortices. Daskopoulos (1989), So (1991), and Petitjeans (1995) have proposed an explanation for the "transition from two to four rolls;" furthermore, most of the numerical and experimental results are given for bends of weak curvature.

The results presented in the Figure 9 have been obtained with a strong curvature ($R/R_0 = 3$), for a Reynolds number $R_e = 1300$ and a Dean number K = 750. The sequence of 12 pictures shows two pairs of vortices, symmetrical with respect to the plane of the bend. The spatial evolution is recorded between 1.4 and 1.5 d downstream of the test zone, the distance between two consecutive planes being 10 mm and the time gap between two recordings being 0.5 s (swirler not used).

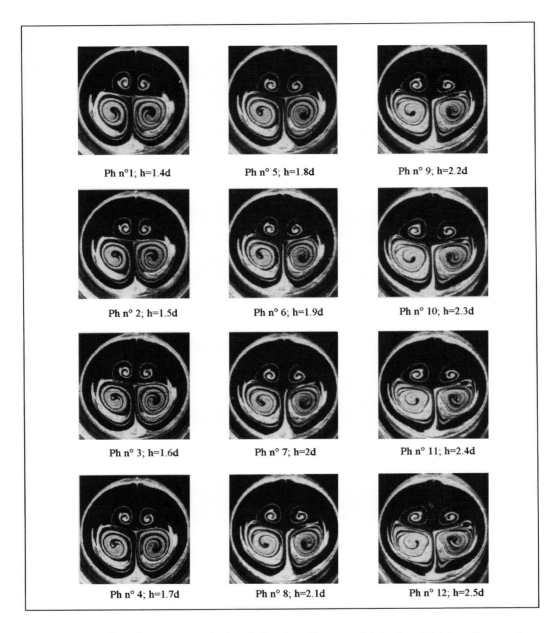

Figure 9 Water flow downstream of a bend. d = pipe diameter, 1.4 d < h < 2.5 d; h = distance from the bend output. Reynolds number, R_e = 1300; Dean number, K = 750; time gap between two pictures, Δt = 0.5 s.

Simultaneous Multiple Tomography

The necessity to study 3-D flows has induced the recent use of mobile sheets with a sequential movement or oscillating sweep (see "Movable Plane Sheets"). Performing in the study of stationary flows, these techniques are of course limited, even unusable, in the case of strongly nonstationary phenomena. In this case the experimenter has to follow the temporal evolution of a sole plane of the fluid, thus losing the space information, or must shift the analysis plane as quickly as possible, losing the strict simultaneity.

Chapter two: Dynamic and multiple laser light sheets

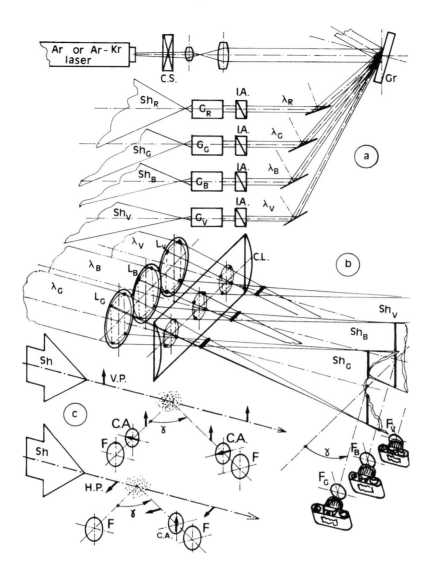

Figure 10 Wavelength coding tomography. C.s. = common shutter, Gr = grating, IA = intensity adjustment, G = sheet generator, Sh = laser sheet, F = interference filters, VP = vertical polarization, HP = horizontal polarization, CA = crossed analyzer. (a) Multiple sheet with chromatic coding; (b) recording configuration; (c) decreasing of the background Rayleigh scattering by polarizing filtering.

Wavelength Coding Tomography

To solve the problem of the strict simultaneity, we have proposed to use a polychromatic tomographic illumination associated with an interference filtering (Prenel et al., 1992; Polidori et al., 1992). A dispersing component, for example, a grating or a prism, is lighted by a multiline laser and allows to create a set of N close laser sheets. During the recording, the same number N of cameras is used. Each of them is focused on one plane, the chromatic selection being assumed by interference filters; so each recording system sees only its own visualization plane. The strict simultaneity is given by a common shutter setup near the laser output (Figure 10).

The most simple configuration consists of using the two powerful lines of a classical argon laser (green, λ_G = 514 nm and blue, λ_B = 488 nm). The power balancing between the two ways is possible by means of a neutral density or by changing an adjustment parameter in one of the sheet generator G. For more sophisticated experiments, more than two wavelengths can be used: with a mixed argon–krypton laser, four visualization planes can be set easily (red, green, blue, and violet) (Figure 10). Of course the complexity and cost of the experiment increase with this number of planes; but another limitation is the overcrowding due to the presence of a set of recording systems. Thus, the use of three wavelengths seems to be a good compromise. Different recording configurations (observation angle γ) can be used for two or three visualization planes (Figure 10b). But the main advantage of this method appears in the case of an orthogonal observation ($\gamma = 0$): though the sheets are completely overlapping for the observer, the interference filtering allows to select precisely a "back" image.

An example is presented in Figure 11. A rectangular foil is moving in a water tank with a constant velocity u and an angle of incidence $\pi/2$ (span A = 25 cm, thickness B = 3 mm, chord C = 5 cm). The selected visualization area is near a foil extremity that is limited by a plane transparent plate. The water is seeded with 100 µm rilsan particles (volumic mass 1.06 g/cm^3) and illuminated by three parallel polychromatic laser sheets; the gap between two sheets is Δz = 2.5 cm.

The time scale t* is dimensionless.

$$t^* = t\frac{u}{C} \tag{17}$$

where t = 0 corresponds to the beginning of the foil movement and the Reynolds number Re_c is related to the chord length C.

Polarizing Coding: Increasing of the Signal-to-Noise Ratio

The signal to be recorded consists of the light scattered by the tracers. If the carrying fluid also induces a scattering phenomenon caused by the presence of submicronic impurities or even due to molecular Rayleigh mode, the corresponding light constitutes a background noise which decreases the image contrast.

The use of a polarized laser allows to free oneself of this background, taking advantage of the results of the light scattering theory. The indispensable condition is the use of micronic tracers (diameter > λ) giving Mie light scattering, easy to separate from Rayleigh light scattering (diameter << λ). The greater part of oil droplets for seeding air flows or spherical solid particles for seeding liquid flows verify this condition.

If we use a laser with a horizontal polarization state, the orthogonal direction of observation ($\gamma = 0$) corresponds to the natural disappearance of the Rayleigh background scattering. For other γ values, this Rayleigh scattering is partially present; so a crossed analyzer introduces a complementary filtering function which allows to increase the signal-to-noise ratio.

If we use a laser with a vertical polarization, the Rayleigh scattering noise contribution is present for all γ values, even if $\gamma = 0$. So the crossed analyzer can be superposed on the interference filtering for any observation angles (Prenel and Porcar, 1992) (Figure 10c).

Conclusion

The study of various flows by means of tomographic visualizations is now frequently used in all fields of fluid mechanics; besides, they have been universally present in the different congresses dealing with fluid experiments for about 10 years. At first their success was due

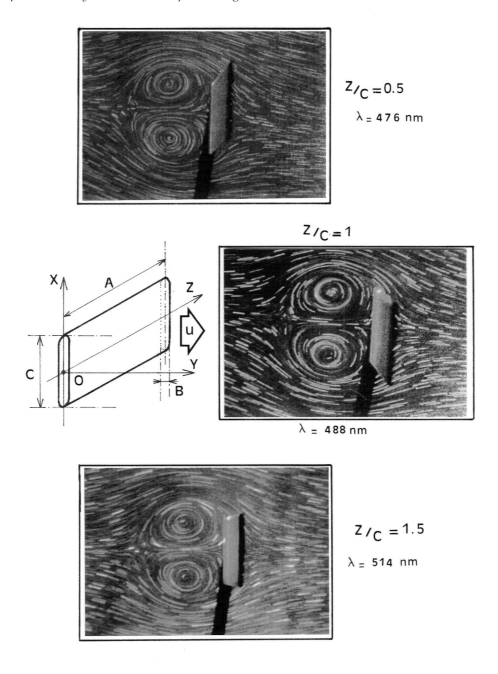

Figure 11 Wavelength coding tomography around a rectangular foil in a water flow (foil velocity u). $Re_c = 1000$, $t^* = t\,u/C = 2$; 1600 ISO emulsion; observation angles $\gamma = 0, -3, +3°$.

to the simpleness of the used optical devices and to the possibility of image recordings through transparent walls whose optical quality is of little importance. Briefly, they are optical methods for experimenters, who are not optics specialists.

The necessity of a temporal analysis has induced the development of beam sweeps or optical oscilloscopes for the investigation of nonstationary phenomena. But it is especially the necessity of 3-D analysis which has induced, during the 1990s, many developments like sheet movements.

However, these methods are often inoperative if the investigated flows are 3-D and at the same time strongly nonstationary. A first response is the resort to simultaneous multiple sheet recordings; thus, the use of a wavelength coding gives interesting results for three or four close planes.

Already we can forecast today the probable evolution of the tomographic techniques during the next years:

- For increasing the beam movements velocity or the sheets shift speed, the electromagnetic scanners can be replaced by acousto-optical deflectors. However, these optical components must still be improved to increase the light power transmission and to avoid distortion or diffraction effects.
- The systematic use of CCD image recorders presents many advantages: flexibility, easy image processings, image control by electronic shutters. Moreover, the connection of a high speed video camera with a high speed optical deflector is a performant combination. But we must not forget that the image resolution is therefore strongly decreased in comparison with a photographic recording. Then we can only hope for rapid progress of the CCD camera resolution.
- Finally, the fast evolution of image processing techniques imposes one to know better the illumination parameters like power density profiles or geometrical configurations. So progress is necessary to master the optical sheet generators better and at the same time to measure the essential sheet parameters: real e^{-2} thickness and width, and characteristic power density. Increasing the quantitative character of these methods must follow. The improvement of the "particle image velocimetry" technique leads to the same conclusion; but this is another story!

References

Ayrault M., Balint J. L., Schon J. P., Measurement of lagrangian velocities in free convection flow by means of a laser visualization system, *Euromech 132*, Lyon, Fance, July 2–4, 1980.

Cabasset P. N., Barbara O., Gajan P., Prenel J. P., Strzelecki A., Visualisation d'écoulements tridimensionnels dans un coude par translation de nappe laser synchronisée, *Visualisation et traitement d'images en Mécanique des Fluides*, St. Etienne, France, June 1995.

Daskopoulos P., Lenhoff A. M., Flow in curved ducts: bifucation structure for stationary ducts, *J. Fluid Mech.*, 203, p. 148, 1989.

Diemunsch G., Prenel J. P., A compact light sheet generator for flow visualizations, *Opt. Laser Technol.*, 19, p. 141, 1987.

Diemunsch G., Prenel J. P., Porcar R., An optical method to the study of instabilities in flows, *Fluid Control and Measurement*, Pergamon Press, Tokyo, September 2–6, 1985.

Kogelnik H., Li T., Laser beams and resonators, *Appl. Opt.*, 5, p. 1550, 1966.

Merzkirch W., *Flow Visualization*, Academic Press, New York, p. 32, 1987.

Petitjeans P., Wesfreid J. E., Deplano V., Vlad G., Effect of curvature on the velocity profile and boundary layer in flow through a curved channel, *La Recherche Aérospatiale*, 0034-1223/95/02, p. 125, 1995.

Philbert M., Boutier A., Méthode optique de mesure des vitesses de particules entraînées dans les écoulements, *Rech. Aerosp.*,1972-3, p. 171, 1972.

Philbert M., Beaupoil R., Faleni J. P., Application d'un dispositif d'éclairage laminaire à la visualisation des écoulements aérodynamiques en soufflerie par émission de fumée, *Rech. Aerosp.*, 1979-3, p. 173, 1979.

Polidori G., Texier A., Coutanceau M., Porcar R., Prenel J. P., Visualisation d'écoulements tridimensionnels par tomographies multiples simultanées, *5ème Colloque National Visualisation et Traitement d'Images*, Poitiers, p. 413, 1992.

Porcar R., Prenel J. P., Robert C., Visualisation d'ondes de choc dans un éjecteur supersonique, *Opt. Commun.*, 14, p. 104, 1975.

Porcar R., Prenel J. P., Diemunch G., Hamelin P., Visualizations by means of coherent light sheets; application to various flows, *Flow Visualization III*, W. J. Yang, ed., p. 123, 1983.

Prenel J. P., Porcar R., Selective polarizing coding laser tomography for aerodynamics, *Opt. Commun.*, 89, p. 12, 1992.

Prenel J. P., Porcar R., Reiniche S., Diemunsch G., Visualisations tridimensionnelles d'écoulements non axisymétriques par balayage programmé d'un faisceau laser, *Opt. Commun.*, 59, p. 92, 1986.

Prenel J. P., Porcar R., El Rhassouli A., Analyse tomographique d'écoulements tridimensionnels par nappes laser en translation, *Opt. Commun.*, 65(2), p. 101, 1988.

Prenel J. P., Porcar R., El Rhassouli A., Three dimensional flow analysis by means of sequential and volumic laser sheet illumination, *Exp. Fluids*, 7, p. 133, 1989.

Prenel J. P., Porcar R., Polidori G., Texier A., Coutanceau M., Wavelength coding laser tomography for flow visualizations, *Opt. Commun.*, 91, p. 29, 1992.

Schneiderman A. M., Sutton G. W., Laser planogram measurements of turbulent mixing statistics in the near wake of a supersonic cone, *Phys. Fluids*, 13, p. 1679, 1970.

So R. M. C., Zang H. S., Lai Y. G., Secondary cells and separation in developing laminar curved pipe flows, *Theor. Comput. Fluid Dyn.*, p. 141, 1991.

Yoda M., Hesselink L., A three-dimensional visualization technique applied to flow around a delta wing, *Exp. Fluids*, 10, p. 102, 1990.

chapter three

Some Typical Mechanisms in the Early Phase of the Vortex-Shedding Process from Particle-Streak Visualization

Madeleine Coutanceau and Gérard Pineau

L.M.F./L.E.A., Université de Poitiers, Poitiers, France

Abstract—This paper illustrates the way the classical technique of particle-streak visualization can be used to provide data on flow topology and, consequently, appears to be a useful tool to bring insight on the phenomena of flow separation and on the successive phases in the vortex formation, development, and shedding process from body walls. Examples of typical mechanisms evidenced by this means, coming essentially from the Laboratoire de Mécanique des Fluides de Poitiers (LMFP), are given considering the early phase of the wake formation behind nominally 2D bluff-bodies of various geometries, impulsively started from rest, either into a single translation or a translation combined with another motion (rotation–oscillation). The structure of the flow at the body junction with towing end plates is also examined.

Introduction

It is widely recognized that visualization is a valuable and even irreplaceable tool to understand complex flows, as those encountered in applications, and to interpret "blind" local probe measurements. This explains why this flow approach encounters an always increasing interest from the researchers of the fluid mechanics community and why new methods, more and more sophisticated, continuously make their appearance in relation to the progress of technology (for a recent presentation and review of the bibliography about these techniques, see, for instance, Freymuth, 1993, and Gad-el-Hak, 1992).

The complementarity of data deduced from the different types of methods is often recommended for a more complete insight of the flows. For example, the recording of the displacement of lit solid particles randomly dispersed in the fluid and captured with a short time of exposure gives a field of streaks (Figure 1[a]) from which the velocity field can be deduced; whereas tracers, such as smoke or dye, emitted continuously from wall, inform us about the convective transfer of vorticity which is produced there (Figure 1[b]). Each principle presents advantages and disadvantages (see Coutanceau and Defaye, 1991, and Gad-el-Hak, 1992).

Although the second technique is now also widely used at the LMFP, the present paper is focused on the first technique, i.e., the particle-streak visualization which presents the advantage of giving the quasi-instantaneous velocity field even if unsteady flows are

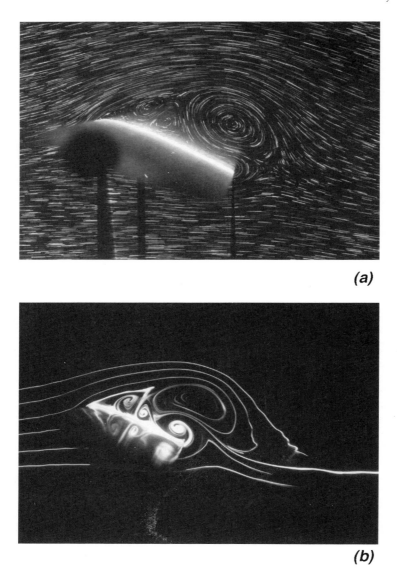

Figure 1 Comparative photographs of the starting flow past a NACA 0012 profile at 20° in incidence, Re = 1000, and t* = 4. (a) Particle-streak visualization; (b) electrochemical dye visualization. (From Meyer, P. and Pineau, G., private communication, 1994.)

examined, provided that a suitable time of exposure with respect to the characteristic time of the flow evolution is used. Under these conditions, and if the flow is quasi-2D, it is possible to reconstruct the streamline patterns from the pictures. However, this reconstruction may present certain ambiguities and consequently induce some errors. This is why it is useful, at a first step, to detect the critical points of the patterns and to identify their type with a view to base the reconstruction on these particular points. It is an approach that we have practiced at the LMFP since the 1980s and which has permitted us to evidence some original phenomena in relation with the formation, development, and shedding of wake vortices, from bluff bodies generating separated flows. Furthermore, this approach facilitates the comparison with data coming from numerical calculations.

Thus, starting from the instream saddles and the surface half-saddles (see next section, "Recalls about Flow Topology"), one determines the singular streamlines (called separatrices) which emanate from (or end to) these critical points following the particle streaks, step by step, while verifying that any streak crosses the curve. This reconstruction, now automatically made on a graphic working station from digitized images (David and Texier, 1994), was made, until now, by hand with a geometrical table. Thus, the corresponding drawn lines, which delimit the vortical structures, permit us to identify these structures and to analyze their time and space evolution. It has been proven that these reconstructions are made with a sufficient precision to serve as a guide in calculations.

Recalls about Flow Topology

It is not the object of the present paper to present a detailed analysis of the concepts on which flow topology is based. For this, the reader is referred, for example, to the reviews by Tobak and Peake (1979, 1982), Perry and Chong (1987), and Chapman and Yates (1991), where a complete bibliography can be found. Here, we will only recall that, according to these concepts, the streamline patterns, in any plane, can be considered in terms of critical or singular points. These points are the particular points where the velocity is zero and the streamline slope is indeterminate, so that, according to Legendre (1977) and based on the assumption that the velocity field is a continuous vector field and can be consequently expanded in terms of space coordinates in any point of the flow, the critical points appear to correspond to the singularities of differential equations and governed by simple kinematical theorems. In this way, three main types of critical points can be distinguished: nodes, saddles, and foci or centers (which are degenerate foci). This list is not at all exhaustive and many other types of degenerate forms of these singularities may be found depending on the order of the velocity series expansion. Furthermore, for a given type of singularity, points within the stream and points located on walls should be distinguished; they correspond to free-slip and no-slip critical points, respectively, according to Perry and Chong (1987). Surface nodes and saddles have been characterized as half-nodes and half-saddles when the streamlines of a two-dimensional slice of the flow cuts a wall. They correspond to separation and attachment points.

In fact, all these points can be classified in only two categories. Nodes, foci, and their degenerate form of centers belong to the same category; each of them is common to an indefinite number of streamlines. Saddles, through which only two streamlines cross each other, constitute the second category (see Figure 2 where separation and attachment points and foci have been differentiated according to whether the streamlines are directed inward toward the point, or outward away from the point considered).

2D-flow streamlines are contained in a plane where the divergence is zero. In consequence, the only critical points that can occur in this plane are saddles and centers. The streamlines that originate from saddles are called sepatrices and divide the flow pattern into distinct regions.

Various rules which the admissible number of simultaneous critical points in a flow pattern should obey have been proposed. For example, in the case of a 2D plane cutting a 3D body, Hunt et al. (1978) found that:

$$\left(\sum_N + \frac{1}{2}\sum_{N'}\right) - \left(\sum_S + \frac{1}{2}\sum_{S'}\right) = -1 \qquad (1)$$

where Σ_N, $\Sigma_{N'}$, Σ_S, $\Sigma_{S'}$ represent the sum of the nodes (including foci and centers), half-nodes, saddles, and half-saddles, respectively. It should be noticed that for a 2D flow, relation (1) reduces to:

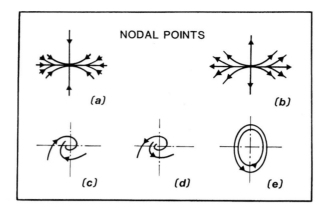

Figure 2 Examples of critical points. (a)(c)(f) Separation points and focus; (b)(d)(g) attachment points and focus; (e) center. (Adapted from Hunt, J. C. R., Abell, C. J., Peterka, J. A., and Woo, H., *J. Fluid Mech.*, 86, 179, 1978.)

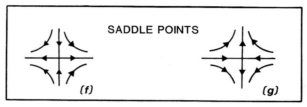

$$\sum_S + \frac{1}{2}\sum_{S'} - \sum_C = 1 \qquad (2)$$

where Σ_C represents the sum of the centers.

These rules, established for steady flows, can be applied generally to quasi-instantaneous velocity fields such as those captured by means of a suitable technique of visualization, in the case of the separated starting flows that we are considering here. Thus, following these rules, the determination of the flow topology can be made with a good reliability, especially for the separatrices which delimit the vortices, the behavior of which we intend to analyze.

However, some care must be taken in this analysis, because the resulting flow structure, which is based on the streamline pattern, is very sensitive to the relative speed of the observer. Indeed, its appearance may completely be changed by modifying this speed, rendering the interpretation sometimes difficult. For example, as mentioned by Lugt (1979), a 2D vortex, which is formed of circular streamlines with center C when viewed in a frame fixed to its axis, appears to be more or less distorted when viewed in a relative translating frame (see Figure 3). Thus, if the relative convection speed V_{rc} of the observer is small with respect to the tangential velocity of the vortex, thus to its strength, it remains formed of closed, however distorted, streamlines and so remains clearly identifiable. Indeed, one can detect its center C and its limiting streamline that now makes a loop which closes in an instream saddle point S. But for a same vortex, the locations of C and S depend on V_{rc}. Thus, if V_{rc} increases, C and S move toward each other, finally merging and giving rise to open streamlines which appear as waves, less and less marked as V_{rc} continues to increase. So the vortex distortion is related both to the difference between the translation speed of the observer and that of the vortex, and to the strength of this vortex. This process is well observed when a vortex detaches from a body; its convection speed, with respect to this body, initially near zero, accelerates along a distance of some body lengths to tend

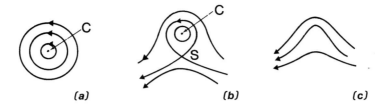

Figure 3 Three aspects of a same vortex according to whether it is fixed or in uniform translation with the relative speed V_{rc} in regard to the observer. (a) $V_{rc} = 0$; (b) V_{rc} is small with respect to the tangential velocity of the vortex; (c) V_{rc} is large. (Adapted from Lugt, H., *Recent Development of Theoretical and Experimental Fluid Mechanics*, Springer-Verlag, New York, 1979, 309.)

progressively toward a limiting value of about 80% of the free stream velocity. In consequence, on a same photograph, vortices can been seen under different shapes; thus it is very important to be clearly precise as to what is the frame of observation.

Thus, viewed in a frame fixed with respect to the body (Figure 4), the separatix limiting a vortex is initially attached to the wall by two half-saddles S' during the phase of formation. Then, when the process of vortex detachment occurs, the two surface half-saddles disappear for the benefit of an instream saddle S_i (i = 1, 2, ...), implying the evolution of the separatrix into a loop closing in S_i. We have shown that the passage from one state to the other arises by means of the transposition phenomenon with a companion contrarotating vortex that will be described in the next section. This loop, closed in S_i, decreases progressively when the vortex is going downstream, under the double action of its relative convection and its viscous diffusion. In consequence, to follow correctly the development of a vortex, the observer should move with its center. But this is not possible exactly when the vortices captured on a same photograph have different velocities. An optimized frame of observation is to be found by seeing the whole pattern with a minimum time shift (Cantwell and Coles, 1983; Perry and Tan, 1957). Furthermore, special attention is needed if noninertial frames are required.

A same type of problem arises when the body wall moves with respect to the observer. Then there are no longer critical points on the boundary since the boundary would no longer have zero relative velocity. In consequence, the vortices appear never attached to the body but always in the form of loops which close in an instream saddle. For example, this case arises when considering the flow pattern generated by a rotating cylinder put in a uniform free stream and observed in a fixed frame, or, what is equivalent, the flow generated by a starting rotating and translating cylinder observed in a frame which accompanies the cylinder only in its translation (see Coutanceau and Ménard, 1985). Thus, it becomes difficult to speak of vortex detachment, according to the previous definition. However, by analyzing the vortex-loop evolution, one can follow the progressive increase of this loop due to the feeding of the vortex in vorticity when it lies in the very near proximity of the body, being quasi at

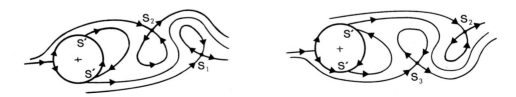

Figure 4 Topological schematics of two successive sequences of the steady vortex-shedding process, the observer being fixed compared to the cylinder. (Adapted from Perry, A. E., Chong, M. S., and Lim, T. T., *J. Fluid Mech.*, 116, 77, 1982.)

rest with respect to the observer. Then its disengagement from this stagnation zone is signaled by the onset of a progressive decrease of the loop.

Some Typical Topological Mechanisms in the Vortex-Shedding Process

By means of such an experimental analysis coming from particle-streak visualization pictures, it has been shown that the vortex-shedding process is topologically based on three main mechanisms, i.e., the vortex splitting, the co-rotating vortex coalescence, and the transposition of the saddles of counterrotating vortices. Shown for the first time in the case of the translating-rotating circular cylinder (see Coutanceau and Ménard, 1985), the third mechanism appears to be necessary for the vortices to disengage from the body wall and is found to be a general process. It has been clearly confirmed by calculation (Badr et al., 1986).

At first, let us explain this transposition phenomenon using the schematics (Figure 5[a]) from Badr et al., (1986) which present the time evolution of the very near wake

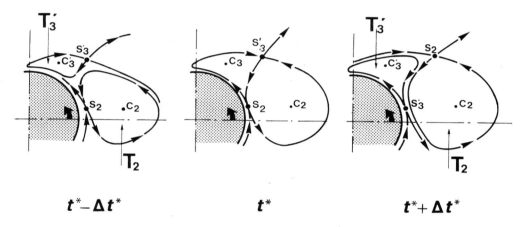

Figure 5 (a)

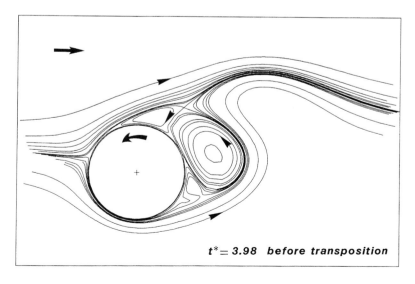

Figure 5 (b)

Chapter three: Vortex-shedding process from particle-streak visualization

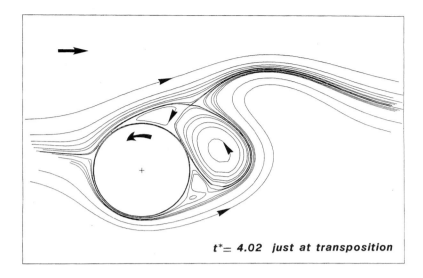

Figure 5 (c)

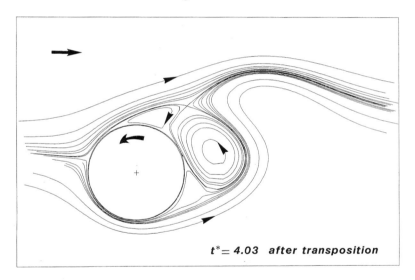

Figure 5 (d)

Figure 5 Transposition process between the closure-saddle points of two counterrotating vortices behind an impulsively started rotating and translating circular cylinder. (a) Schematics deduced from experiment ; (b) streamline patterns numerically generated for Re = 500 and α (speed ratio) = 0.5. (Adapted from Badr, H., Coutanceau, M., Dennis, S. C. R., and Ménard, C., *C. R. Acad. Sci.*, Gauthier-Villars, Paris, 302 (Ser. II, no. 18), 1127, 1986.)

generated by an impulsively started translating–rotating cylinder, at a certain stage of its initial development. The first vortex T_1 has already been shed and is omitted on the drawing. It is seen that, at the time $t^* - \Delta t^*$, the positive vortex T_2, whose center is C_2, is initially closed in the saddle S_2 toward the rotating wall of the cylinder. Whereas, the vortex T_3', whose center is C_3' is closed in S_3' toward downstream. At the subsequent time t^*, the boundaries of the two vortices, which have grown with time, come into contact such as a same streamline joins the two saddles S_2 and S_3'. However, it has been noticed since then

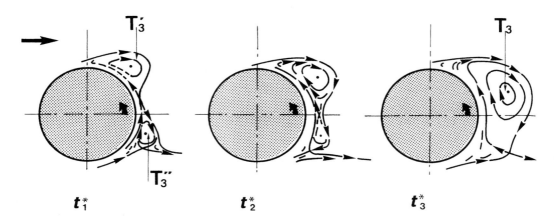

Figure 6 Schematics of the coalescence process of two co-rotating vortices. (From Badr, H., Coutanceau, M., Dennis, S. C. R., and Ménard, C., *C. R. Acad. Sci.*, Gauthier-Villars, Paris, 302 (Ser. II, no. 18), 1127, 1986.)

by Perry and Chong (1987), this configuration is unstable and cannot persist: "It occurs only at an instant during a bifurcation process." Indeed, at $t^* + \Delta t^*$, the two vortices separate again each from the other, but after having echanged their closure saddles. Thus, from this time, the closure point of T_2 is directed toward the free stream and T_2 becomes able to disengage from the very near wake. This is what has been shown in a more complete sketch of the time evolution of the flow pattern deduced from the experiments (see Coutanceau and Ménard, 1985): T_2 will effectively be shed just after this event. Figure 5(b) shows that this transposition phenomenon has been exactly captured by means of calculation, whereas in the experiment, it is too much ephemeral to be captured except by chance. This example confirms the efficiency of developing the experiment and calculation in parallel.

Once again, let us refer to the schematics by Badr et al. (1986) to explain the coalescence of two corotating vortices (Figure 6). Thus, starting at time t_1^* from the two separate vortices with T_3' and T_3'' having their own center and saddle, they will come into contact at the subsequent time t_2^* because of their time-growing, provoking a unique outer separatrix to pass around the two vortices and another inner eight-shaped one to pass around the two centers. The configuration will still have two saddles and two centers. Then, with increasing time, the two cores draw nearer each other, come into contact, and merge at time t_3^* into a single core, inducing the coalescence of the two initial vortices.

It may be remarked that, taking into account the existence of the upstream saddle point, all the proposed schematics are compatible with the law proposed by Hunt et al. (1978) given in "Recalls about Flow Topology". In the coalescence process, one center has disappeared, but simultaneaously one saddle has also disappeared.

Finally, with regard to the vortex splitting phenomenon, it consists of the "reversing" of the process of coalescence, namely a unique vortex gives rise to two corotating vortices. This is generally provoked by the stretching of the vortex under the shear stress of opposite surrounding currents. As an example, see the flow time-evolution in Figure 7, where the inner part of the first shed vortex splits into two cores having two centers surrounded by an eight-shaped separatrix. Then, if the shearing processs is sufficiently strong, the two cores separate from each other and get their own saddle and own center by means of a transposition phenomenon and follow their own trajectories (possibly even in opposite directions as on Figure 7). This splitting phenomenon, also mentioned by Freymuth (1985) and Freymuth et al. (1985), has been found in many situations. As an example, the three topological mechanisms described above are all encountered within a lapse of time as short

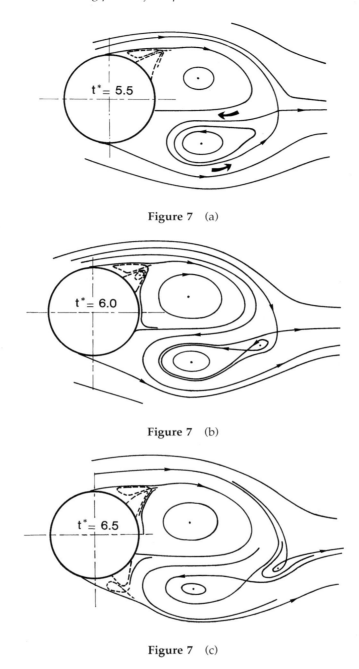

Figure 7 (a)

Figure 7 (b)

Figure 7 (c)

Figure 7 Schematics of the splitting process, in the initiation of the vortex shedding behind a circular cylinder, at Re = 1000. (From Boyce, Ph., Thesis of the University of Poitiers, France, 1986.)

as $4.0 \leq t^* \leq 5.5$ in the case of the impulsively started flow around a circular cylinder fitted with a thin flap positioned at 45° downstream (see Figure 8).

Another phenomenon also appears clearly from the pictures, issued from a particle-streak visualization, showing a slice of quasi-2D flow. This is the change from a 2D to a 3D structure of a vortex. Thus, the schematics of Figure 9 show that a cross section of a pair of 2D vortices, such as those encountered in the initial phase of an attached symmetrical recirculating zone, will give streamlines enclosing centers, while in the case of a

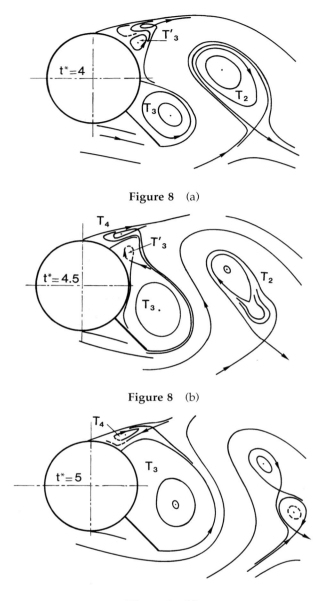

Figure 8 (a)

Figure 8 (b)

Figure 8 (c)

3D-recirculating zone, spirally shaped, the same flow slice will present foci instead of cores. Thus, the detection of the change of instream centers into foci indicates the appearance of 3D effects in starting 2D-recirculating flows. This phenomenon is often observed earlier than any other typical 3D manifestations such as the crossing of the studied plane by the particles. Indeed, as shown by Pineau et al. (1992), and since then by Polidori (1994), the vortex cores are perturbed at the first, especially when the 3D perturbations come from the body end extremities (end plates, for example). The limitation in length of the body provokes a spirally spanwise current which renders the wake vortices effectively three-dimensional (Polidori et al., 1992).

Now, referring to selected examples, it will be shown how these particular phenomena arise effectively in the vortex-shedding process. We will consider, in the section "Time Evolution of the Topological Pattern in the Median Section of the Flow around a Nominally

Chapter three: Vortex-shedding process from particle-streak visualization

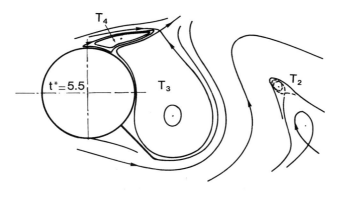

Figure 8 (d)

Figure 8 Splitting, coalescing, and saddle-transposition processes in the vortex shedding from a starting circular cylinder fitted with a flap. (From Boyce, Ph., Thesis of the University of Poitiers, France, 1986.)

2D Bluff Body", successively a central cross section of several nominally 2D flows with body wall fixed with respect to the observer or in motion relative to it, and finally, in the section "Topological Pattern Example of a Wall Junction Flow", we will give an example of a junction surface flow where a greater variety of critical points may be encountered.

Technique for Obtaining Pictures

Before presenting examples of the topological results under consideration here, we will first recall how the pictures, from which these results are extracted, are obtained at the LMFP. Thus, the experimental conditions being selected, the time evolution of the flow pattern is deduced from sequences of photographs taken, in a same run, by a camera which follows the studied nominally 2D model in a vertical translation inside a tall tank filled with a transparent liquid (Figure 10); water and different transparent oils are chosen in adequacy with the range of Reynolds number under investigation.

For this, the model is attached by means of a system of end plates and rods to a very heavy carrier and towed, with a constant speed V_0, downward or upward, according to the experimental setup; three setups are available at the present time at the LMFP. In each setup, the motion of the body-frame system is induced by the combination of the gravity force and of the retarding force of a dashpot; the start is made quasi-instantaneous. The precise value of V_0 ($2 \text{ cm/s} \leq V_0 \leq 20 \text{ cm/s}$) is obtained by means of complementary balance

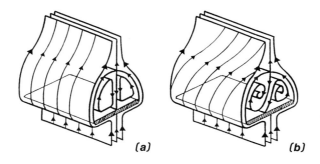

Figure 9 Spanwise vortex pair forming the initial recirculating zone of a flat plate normal to the oncoming stream; schematics of successive structures. (a) 2D structure, (b) 3D structure. (From Polidori, G., Thesis of the University of Poitiers, France, 1986.)

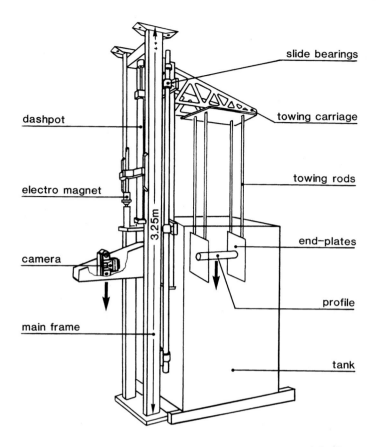

Figure 10 Schematic of one of the LMFP experimental facilities.

weights and known by means of an electronic counter associated with a device including diodes regularly positioned along the height of the fixed tank and a photo cell cotranslated with the body outside the tank. The sections of the present tanks are rectangular and may be modified by addition of vertical plates.

The flow is made visible by seeding the liquid of the tank with dispersed diffusing tracers: small white particles of rilsan are the most often used because of their suitable density, their thick shape (almost sphere) avoiding preferential reflection of the incident light, and because of their chemical inertness. These particles are lit by a light sheet coming either from a powerful arc projector (120 A, 180 V) or an argon laser (5 and 10 W). The flow-recording camera is put on a platform fixed to the translating carrier.

To get the flow with sufficient details, the sequential recording is made at sufficient small time intervals and the experiments are carried out with bodies of sufficient size D (D is usually about 5 cm, but can be increased up to 15 or even 20 cm). However, the period of the normalized time t^* ($t^* = tV_0/D$) during which the flow development can be observed, as well as the aspect ratio of the model, is limited respectively by the height H_t and the width W_t of the tank, and this all the more as the body is bigger: $t^*_{MAX} < H_t/D$ and $AL < W_t/D$. Besides, the flow is more or less confined by the transversal dimension l_t of the tank, generating another parameter which can become significant, i.e., the blockage ratio $\lambda = D/l_t$, especially for low Reynolds number regime.

The three LMFP tanks are $T_{a1} = W_t \times l_t \times H_t = 100 \times 80 \times 150$ cm^3, $T_{a2} = 56 \times 46 \times 146$ cm^3, $T_{a3} = 56 \times 46 \times 100$ cm^3, authorizing, because of the limiting length of their associated

Chapter three: Vortex-shedding process from particle-streak visualization 55

dashpot, a maximum vertical displacement of 80, 80, and 40 cm, respectively. Each of them are equipped with the same type of mechanism for the generation of the translation motion, except that T_{a1} and T_{a2} work downward, whereas T_{a3} works upward (see Coutanceau and Bouard, 1977). Furthermore, these setups can be equipped with complementary mechanisms to also make the model rotating (see Coutanceau and Ménard, 1985) or oscillating (see Ohmi et al., 1990) during the translation, the superposed motions being started together impulsively.

The spanwise structure of the flow can be viewed by turning the model support through 90°. However, in addition, two of the experimental setups are also equipped with devices authorizing a more specific study of the 3D effects, offering the possibilities to capture the flow in regularly spaced cross sections, all along the span of the model. One of the possibilities is to translate the laser support along a horizontal rail and to position it, with great precision, in front of the selected cross section (Pineau, 1992). The more recent one is based on the simultaneous capturings of the flow, in three different cross sections, lit by three light sheets of different colors (green, blue, and violet). The simultaneous recordings are carried out by three synchronized cameras equipped with suitable filters; for example, each of them sees only one of the three planes. This technique has been elaborated in concert with Prenel et al. (1992); it is described in the present book by Prenel et al. (1997). This is a very efficient and reliable tool for investigating such highly unsteady flows such as those resulting from an impulsive start.

Time Evolution of the Topological Pattern in the Median Section of the Flow Around a Nominally 2D Bluff Body

In this section, we will distinguish the case of starting flows around fixed and moving bodies.

Body Walls Fixed to the Observer

As examples of starting flows around bodies whose walls are fixed with respect to the frame of observation, we will consider successively the fundamental symmetrical case of the circular cylinder and subsequently the asymmetrical case when the circular cylinder is fitted with a thin flap positioned at 45° from the downstream axis. This latter case will be related to that of a flat plate, schematizing a thin profile, set with 30° angle compared to the oncoming free stream. The selected Reynolds number Re is 1000, or near 1000, and topological schematics and photographs are presented in parallel. In addition, some complements are given for Re = 4500 for a detailed examination of the secondary vortex behavior.

Thus, Figure 11 presents some stages of the time evolution of the central cross-sectional pattern of the flow around a circular cylinder, whose aspect ratio AL is 10 and which is towed by two large, parallel end plates. It is shown how the two main vortices grow behind the symmetrical body, at first as a symmetrical pair (Figure 11[a]), then become asymmetrical with the detachment from the wall of one of the vortices which turns downstream (Figure 11[b]) under the infiltration of the free stream and consequently becomes ready to be shed. When time increases (Figure 11[c]), this vortex (upper vortex on the present sequence of pictures) stretches downstream losing its downstream part, while a spanwise current coming from the end plate (Pineau et al., 1992) makes the vortex centers change into spirally foci. Then the clear splitting and shedding of this vortex downstream part occurs (which plays the role of the airfoil starting vortex), while the upstream part is going back upstream (Figure 11[d]). It will reattach to the cylinder wall at a slightly later stage by means of a transposition of its closure saddle point with the two half-attachment saddle points of the lower vortex, which becomes ready to be shed in its turn (see Boyce, 1986).

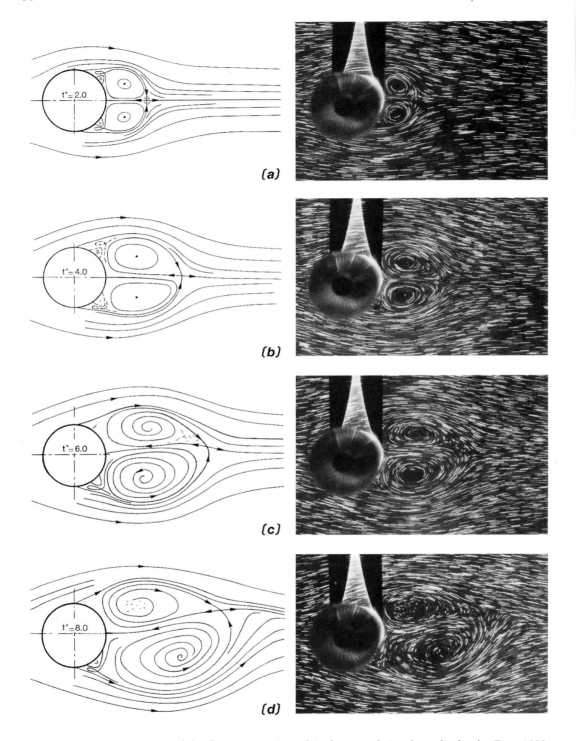

Figure 11 Time sequences of the flow past an impulsively started circular cylinder for Re = 1000. (From Pineau, G., Thesis of the University of Poitiers, France, 1992.)

The interrupted lines indicate part of the flow where the particles either have very low velocities, or cross the illuminated plane because of the three-dimensionality of the flow, or are not clearly visible.

Chapter three: *Vortex-shedding process from particle-streak visualization*   57

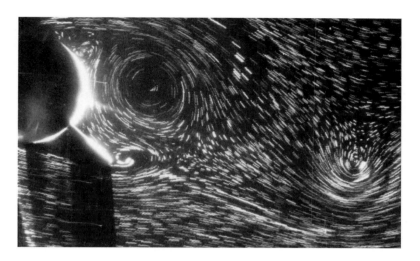

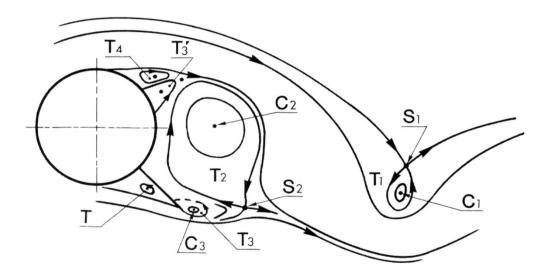

Figure 12   A time stage of the wake pattern behind an impulsively started circular cylinder fitted with a flap: Re = 1000, t* = 3. (From Boyce, Ph., Thesis of the University of Poitiers, France, 1986.)

Pineau (1992) showed that the vortex-shedding mechanism remains roughly of the same type for a shorter cylinder with AL = 5 and in the other cross sections. However, as the cross-plane is shifted toward the end plates, the 3D effects interact sooner, the recirculating zone becomes shorter and thinner, and the downstream movement of the vortex cores is clearly slowed down. Consequently, the asymmetry is less marked, the vortex cores remain in the near-wall proximity, and the onset of the vortex shedding is delayed; its intensity is weakened. This is in quite agreement with the findings of Perry and Steiner (1987) related to the flow behind a normal flat plate in quasi-established regime: "The cavity vortices appear to be confined closer to the plate than when no endplates were fitted."

In Figure 12, we have chosen to show the flow behind a circular cylinder fitted with a flap, at the time-stage t* = 3 (so the body has traveled downward along a distance of three times its diameter), the topological configuration appearing to be quite interesting. Indeed, it implies different types of vortices: the traveling and consequently distorted starting vortex T_1, the newly shed vortex and consequently still rounded shaped T_2, the forming trailing vortex T_3 attached to the flap, the secondary vortex T_3' attached to the cylinder wall and originating from a secondary flow separation, the secondary vortex T_4 coming from the splitting of T_2 under the increase of T_3' and the vortex T attached to the lower side of the flap. At a later stage (see Boyce, 1986), T_3 and T_3', which have grown, coalesce to give rise to a unique positive vortex T_3. Subsequently, T_3 will be shed in its turn due to the transposition of its two wall half-saddles with the closure saddle of T_4.

Compared with the starting flow behind a circular cylinder without flap, within similar conditions (i.e., Re = 1000, AL = 10), it is seen that the shedding process is highly accelerated and the intensity of the starting vortex highly enhanced. For a more complete analysis, see the work by Boyce (1986) where the Reynolds number has been increased up to 4500 and the influence of polymer additive has been evaluated (see also Figure 4.21 in Boger and Walter, 1993).

Another example of initial asymmetrical flow is shown in Figure 13, which presents five stages of the starting flow pattern around a thin flat plate, inclined at 30° compared to the oncoming stream (from Polidori and Coutanceau, 1992). Interesting processes are evidenced such as the splitting of the initial leading-edge vortex T_2 because of a secondary separation (T_3'), then the coalescence of the vortices T_3 and T_3', the saddle transposition between T_4 and the combined vortex T_3–T_3', and finally the shedding of the trailing-edge vortex T_3 giving place for the leading-edge vortex T_4 to grow.

Higher-order separation phenomena are more precisely shown in Figure 14, giving details of some time stages of the flow, in the wall-separation area, in the case of the circular cylinder at Re = 4500 (Coutanceau et al., 1988). The visualization reveals clearly a secondary separation, but also a tertiary separation, as well as the periodic increase and decrease of the corresponding vortices, especially that of the positive secondary vortex, inducing its intermittent opening on the separated shear layer (see Figure 14[d]) and consequently the periodic shedding of small amounts of rotating fluid (Coutanceau and Defaye, 1991). This is well in agreement with the recent results by Nagata et al. (1989). The early appearance of a 3D structure in the form of a helicoidal motion (see Figure 14[c]) and the disorganization of the tertiary vortices due to a transversal motion of the fluid and to the first manifestation of turbulence (Figure 14[d]) is also evidenced. These secondary instabilities have been shown to rapidly enhance the instability of the whole wake itself.

Body Walls in Relative Motion to the Observer

Let us show now some typical topological behaviors when the body is in motion relative to the observer. We will present successively the case of a rotating circular cylinder and of an oscillating ellipse.

Figure 15 presents the way a frontal separation takes place, in the very early stage of the flow development past a circular cylinder (R) impulsively started both into uniform translation (V_0) and rotation (ω) at Re = 1000, when the rotating-to-translating speed ratio ($\alpha = \omega R / V_0$) is sufficiently high ($\alpha = 3$ in Figure 15). The computer-generated streamline patterns (Badr et al., 1990) are in excellent agreement with the visualization, except that from $t^* = 2.0$, a 3D phenomenon (spiral rolling up) arises in the experiment from the time where the frontal vortex is opening upstream.

Figure 16 establishes a comparison, for the same rotating and translating circular cylinder, at the time-stage $t^* = 3$, when Re = 1000 and 3000 and $\alpha = 0.5$ and 1.0, respectively. The schematics deduced from the pictures explain the topological configuration of the primary and secondary vortices and show the respective influence of the speed ratio α and of the Reynolds number Re. In particular, it is shown that the simultaneous increase of Re and α provokes the shedding of the lower vortex before that of the upper vortex (see Figure 16[d]), contrary to what happens for smaller Reynolds numbers, taking into account the

Chapter three: Vortex-shedding process from particle-streak visualization 59

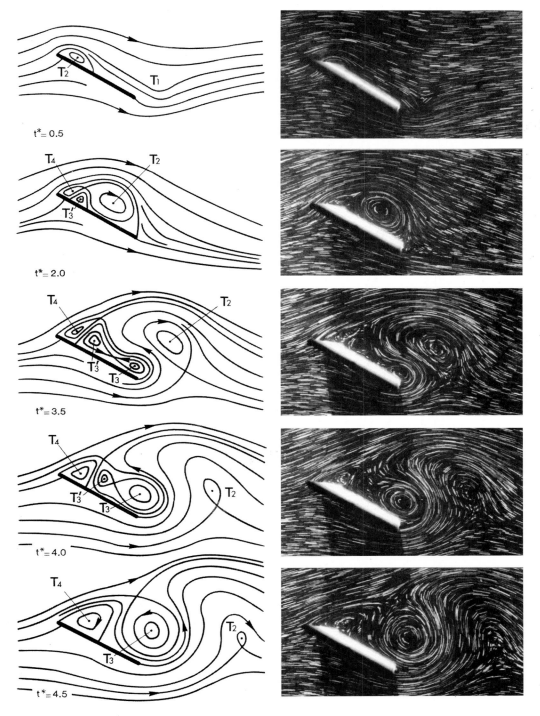

Figure 13 Time sequences of the flow past an impulsively started flat plate, at 30° incidence, for Re = 650. (From Polidori, G. and Coutanceau, M., 12th Congress Français de Mécanique, Burnage, H., Ed., Strasbourg, 1992.)

relative sign of the rotation (Coutanceau and Ménard, 1985). This apparent anomaly is due to the progressive slowing down of the near-wake development in the case of the translating cylinder (forwake phenomenon evidenced by Bouard and Coutanceau, 1980) when Re

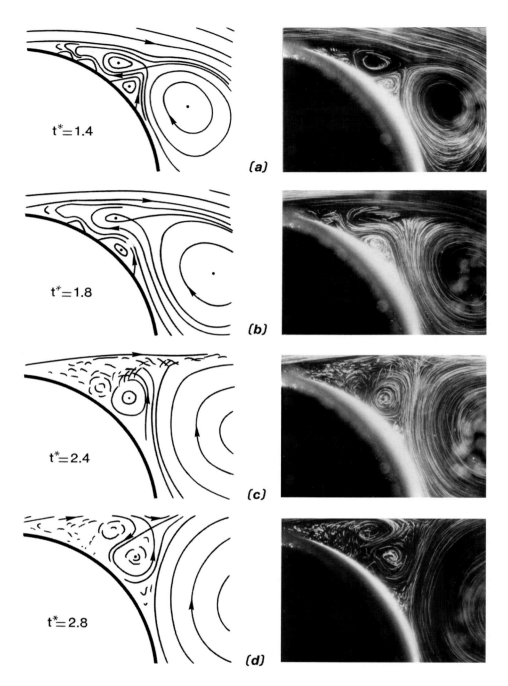

Figure 14 Higher-order separation phenomena for an impulsively started circular cylinder at Re = 4500. (From Coutanceau, M., Boyce, Ph., and Guérineau, G., *C.R. Acad. Sci.*, Gauthier-Villars, Paris, 306 (Ser. II), 1259, 1988.)

increases, implying a relative decrease of the rotation effect on the cylinder side in opposite motion relative to the free stream, because of the corresponding higher local Reynolds number (see Li, 1990). This evolution clearly appears when comparing Figures 16(a) and (c) and 16(b) and (d). The schematics of Figure 17 explain the three shedding modes that have been experimentally observed by Li and Coutanceau (1992) when Re and α vary.

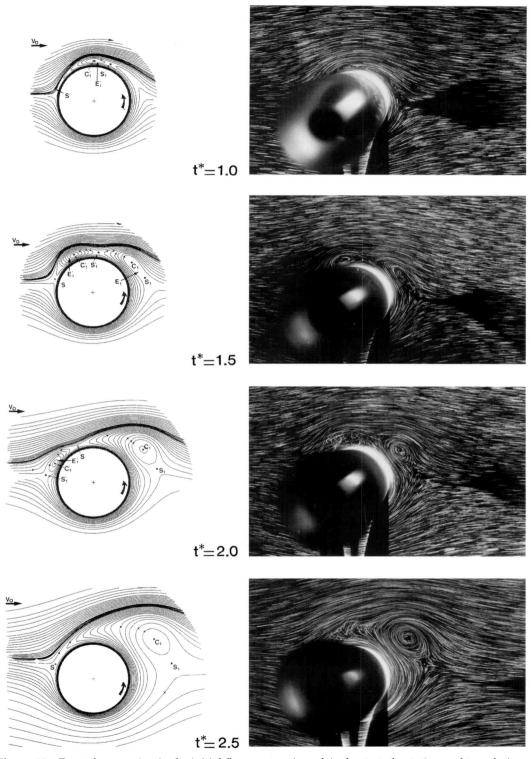

Figure 15 Frontal separation in the initial flow past an impulsively started rotating and translating circular cylinder at Re = 1000, α = 3. Left: numerical calculations; right: experiments. (From Badr, H., Coutanceau, M., Ménard, C., and Dennis, S. C. R., *C.R. Acad. Sci.*, Gauthier-Villars, Paris, 310 (Ser. II), 101, 1990.)

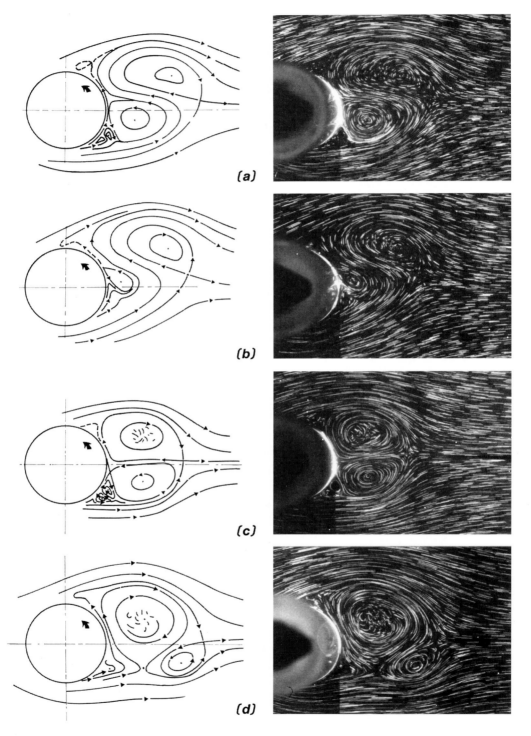

Figure 16 Flow patterns past an impulsively started rotating and translating circular cylinder at $t^* = 3$. (a) Re = 1000, α = 0.5; (b) Re = 1000, α = 1.0; (c) Re = 3000, α = 0.5; (d) Re = 3000, α = 1.0. (From Li, Z., Report of Diplôme d'Etudes Approfondies of the University of Poitiers, France, 1990.)

Chapter three: *Vortex-shedding process from particle-streak visualization*   63

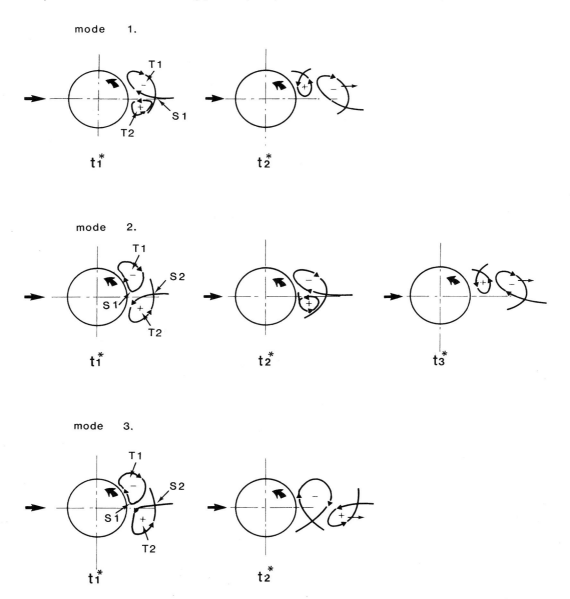

Figure 17 Topology of the three possible modes in the initiation of the vortex shedding behind an impulsively started rotating and translating circular cylinder. (From Li, Z. and Coutanceau, M., *Visualisation et Traitement d'Images en Mécanique des Fluides,* Coutanceau, M. and Coutanceau, J., Eds., Poitiers, 1992, 275.)

These results are in agreement with the calculations by Chang and Chern (1991), except some shifting in the Re and α values for which the phenomena arise.

Some typical flow patterns, induced by a thin ellipse (1/10) suddenly subjected to a translation combined with a periodic pitching, are presented in Figure 18, for Re = 3000 (Re is based on the greater axis c) and a dimensionless frequency f^* ($f^* = fc/V_0$) of 0.5. Once again, a frontal separation is observed at the leading edge of the ellipse (Figures 18[a] and [b]). In addition, the accumulation of the leading edges' vortices on the upper surface of the ellipse is clearly evidenced (Figures 18[b] and [c]). It is due to the relatively high effect of the rotation compared to the translation; the streamline patterns come from numerical calculations (see Ohmi et al., 1990).

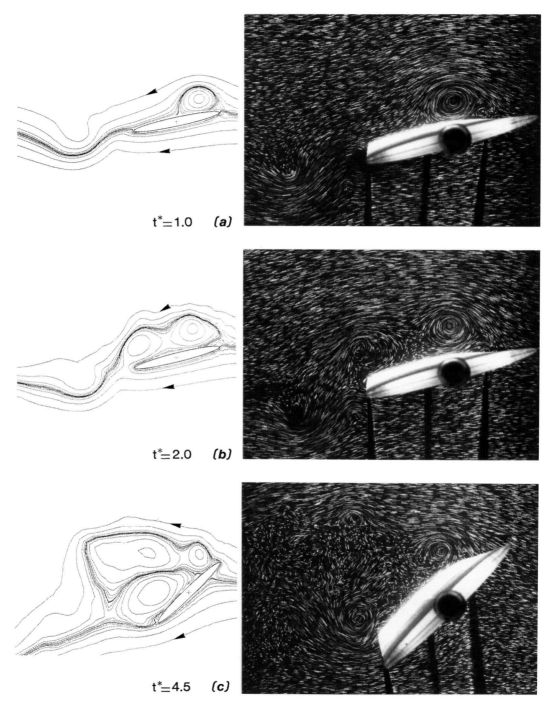

Figure 18 Time sequences of the starting flow past a thin oscillating in pitch and translating elliptic cylinder for Re = 3000 and f* = 0.5. Left: numerical calculations; right: experiments. (From Ohmi, K., Coutanceau, M., Ta Phuoc Loc, and Dulieu, A., *J. Fluid Mech.*, 211, 37, 1990.)

Topological Pattern Example of a Wall Junction Flow

Having evidenced that 3D phenomena appear progressively with time when analyzing the wake-in-formation topology behind quasi-2D bodies (for example, vortex centers evolve into foci), a systematic exploration of the flow in cross sections evenly spaced along the body span (see the section, "Technique for Obtaining Pictures"), and in the spanwise section reveals that these 3D effects mainly arise from spanwise currents taking their origin in the separated boundary layer of the end plates which are used to tow the body in the tank, at least in the limit of the lapse of time under consideration in the LMFP experiments (Pineau et al., 1992, and Polidori and Coutanceau, 1995). Consequently, the flow which develops along each towing end plate has been visualized and topologically interpreted. Figure 19(a) presents an example of the end-wall flow pattern obtained for Re = 650, at time $t^* = 4.5$, when the towed body is a normal flat plate (taken as a schematization of thin profile). The interpreted topology (Figure 19[b]) appears to be composed of a greater variety of singular points than the flow patterns of the central instream slices considered above. The detection and identification of these singular points and the reconstruction of the corresponding singular streamlines evidence clearly, once again, the flow structure and permit us to follow its development with time and its evolution with Reynolds number.

Thus, following the trajectories of the singular points, it has been shown that the horseshoe configuration sets up suddenly in a bifurcation process, at $t^* \approx 1.5$, after the abrupt separation from the wall of the upstream stagnation point S_1, which induces the separation line Ls_{am} and the complementary node N_1 and attachment line La_{am}. The near wake of the flat plate is composed of two main roll-ups around the foci F_1 and F_2 and, more downstream, of a singular attachment line La_{av}, which is the footprint of the outer boundary of the recirculating zone of the near cross-flow. It has been found that the 3D spanwise currents, mentioned above, come from these foci F_1 and F_2.

It may be remarked that, at this time $t^* = 4.5$, the upstream part of the junction flow is not very far from being established (Polidori and Coutanceau, 1995), whereas the downstream part is still in a transient phase and becomes oscillatory and highly unstable from $t^* \approx 7$.

A similar configuration has been found when Re = 1000. But at Re = 3000, the structure of the flow is still more complex (Polidori, 1994).

Conclusion

Basing the analysis on general topological rules and after having recalled some limitations and the need of some precautions, we have shown how images issued from a particle-streak-visualization technique can be used as a valuable tool to interpret flow configuration and consequently constitute a very useful data base to understand the fundamental mechanisms of flow development.

In the present paper, the interest has been focused on the wake formation and the initiation of the vortex-shedding process behind nominally 2D bodies. Starting flows around a circular cylinder fitted or not with a flap and around a NACA profile in incidence, impulsively subjected to a single translation or to a translation combined with a rotation or a pitching oscillation, and captured by an accompanying translating camera, have been successively considered as typical examples.

It has been shown that the same basic mechanisms of secondary separations — vortex splitting, coalescing, saddle transposition — are encountered whatever the body and its motion. However, specific configurations, such as an apparent frontal separation, have been pointed out when the body is in motion compared to the observer. Inversion in the

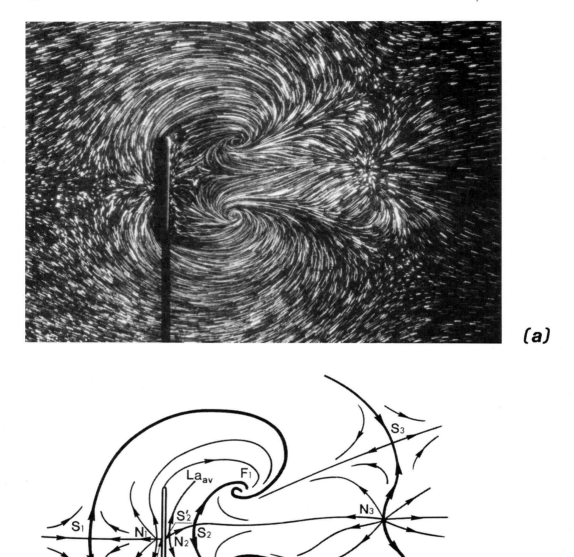

Figure 19 Wall-flow pattern at the junction of a normal flat plate with an end plate at $t^* = 4.5$, for $Re = 650$. (From Polidori, G. and Coutanceau, M., *Visualisation et Traitement d'Images en Mécanique des Fluides*, Schon, J.-P., Ed., Saint-Etienne, 1995.)

initial vortex-shedding order, or vortex accumulation have also been evidenced in the case of the combined motions. Finally, the particularities of the topology of the end-wall junction flow have been presented.

This type of analysis appears susceptible to be rich in new data and of promise. Especially, this approach is probably a reliable way to guide flow analysis by means of automatic processing of digitized images.

Acknowledgments

The authors are very grateful to Dr. Jean-René Defaye who kindly helped them in checking English imperfections in their manuscript, and Dr. Patrick Hébrard for his relevant suggestions.

References

Badr, H., Coutanceau, M., Dennis, S. C. R. and Ménard, C., Sur les phénomènes de transposition et de coalescence de tourbillons dans les écoulements décollés, *C.R. Acad. Sci.*, Gauthier-Villars, Paris, 302 (Ser. II, no. 18), 1127, 1986.

Badr, H., Coutanceau, M., Dennis, S. C. R. and Ménard, C., Unsteady flow past a rotating circular cylinder at Reynolds number 10^3 and 10^4, *J. Fluid Mech.*, 220, 459, 1990.

Badr, H., Coutanceau, M., Ménard, C. and Dennis, S. C. R., Effets particuliers dus à la rotation dans les phénomènes de décollement autour d'un obstacle au démarrage, *C.R. Acad. Sci.*, Gauthier-Villars, Paris, 310 (Ser. II), 101, 1990.

Boger, D. V. and Walter, K., *Rheological Phenomena in Focus*, Elsevier, New York, 1993, 98.

Bouard, R. and Coutanceau, M., The early stage of development of the wake behind an impulsively started-cylinder for $40 < Re < 10^4$, *J. Fluid Mech.*, 101, 583, 1980.

Boyce, Ph., Effets d'additifs polymèriques sur la formation des sillages, Thesis of the University of Poitiers, France, 1986.

Cantwell, B. and Coles, D., An experimental study of entrainment and transport in the turbulent near wake of a circular cylinder, *J. Fluid Mech.*, 136, 321, 1983.

Chang, C.-C. and Chern, R.-L., Vortex shedding from an impulsively started rotating and translating circular cylinder, *J. Fluid Mech.*, 233, 265, 1991.

Chapman, G. T. and Yates, L. A., Topology of flow separation on three-dimensional bodies, *Appl. Mech. Rev.*, 7, 329, 1991.

Coutanceau, M. and Bouard, R., Experimental determination of the main features of the viscous flow in the wake of a circular cylinder in uniform translation. I. Steady flow, *J. Fluid Mech.*, 79, 231, 1977.

Coutanceau, M. and Defaye, J.-R., Circular cylinder wake configurations: a flow visualization survey, *Appl. Mech. Rev.*, 44 (no. 6), 225, 1991.

Coutanceau, M. and Ménard, C., Influence of rotation on the near-wake development behind an impulsively started circular cylinder, *J. Fluid Mech.*, 158, 399, 1985.

Coutanceau, M., Boyce, Ph. and Guérineau, G., Sur le rôle des décollements d'ordre supérieur sur la naissance de l'instabilité secondaire de la frontière du sillage d'un cylindre circulaire, *C.R. Acad. Sci.*, Gauthier-Villars, Paris, 306 (Ser. II), 1259, 1988.

David, L. and Texier, A., Application of a particle streak velocimetry technique to unsteady flow study, *Flucome' 94*, Hébrard, P., Ed., Toulouse, 1994, 515.

Freymuth, P., The vortex patterns of dynamic separation: a parametric and comparative study, *Prog. Aerospace Sci.*, 22, 161, 1985.

Freymuth, P., Flow visualization in fluid mechanics, *Rev. Sci. Instrum.*, 64 (no. 1), 1, 1993.

Freymuth, P., Bank, W. and Palmer, M., Further experimental evidence of vortex splitting, *J. Fluid Mech.*, 152, 289, 1985.

Gad-el-Hak, M., Splendor of fluids in motion, *Visualisation et Traitement d'Images en Mécanique des Fluides*, Coutanceau, M. and Coutanceau, J., Eds., Poitiers, 1992, 27.

Hunt, J. C. R., Abell, C. J., Peterka, J. A. and Woo, H., Kinematical studies of the flows around free or surface-mounted obstacles; applying topology to flow visualization, *J. Fluid Mech.*, 86, 179, 1978.

Legendre, R., Lignes de courant d'un écoulement permanent: décollement et séparation, *La Rech. Aerosp.*, 6, 327, 1977.

Li, Z., Développement du proche sillage d'un cylindre circulaire démarrant impulsivement en rotation et en translation pour $10^3 < \text{Re} < 10^4$, Report of Diplôme d'Etudes Approfondies of the University of Poitiers, France, 1990.

Li, Z. and Coutanceau, M., Développement initial du proche sillage d'un cylindre circulaire démarrant impulsivement en rotation et en translation pour $10^3 < \text{Re} < 10^4$, *Visualisation et Traitement d'Images en Mécanique des Fluides,* Coutanceau, M. and Coutanceau, J., Eds., Poitiers, 1992, 275.

Lugt, H., The dilemma of defining a vortex, *Recent Development of Theoretical and Experimental Fluid Mechanics,* Springer-Verlag, New York, 1979, 309.

Meyer, P. and Pineau, G., private communication, 1994.

Nagata, H., Nagase, I. and Ito, K., Unsteady flows past a circular cylinder started impulsively in the Reynolds number range $500 < \text{Re} < 10000$, *Bull. J.S.M.E.,* 32 (no. 4), 540, 1989.

Ohmi, K., Coutanceau, M., Ta Phuoc Loc, and Dulieu, A., Vortex formation around an oscillating and translating airfoil at large incidences, *J. Fluid Mech.,* 211, 37, 1990.

Perry, A. E. and Chong, M. S., A description of eddying motions and flow patterns using critical-point concepts, *Annu. Rev. Fluid Mech.,* 19, 125, 1987.

Perry, A. E. and Steiner, T. R., Large-scale vortex structures in turbulent wakes behind bluff bodies. I. Vortex formation processes, *J. Fluid Mech.,* 174, 233, 1987.

Perry, A. E. and Tan, D. K. M., Simple three-dimensional vortex motions in coflowing jets and wakes, *J. Fluid Mech.,* 141, 197, 1957.

Perry, A. E., Chong, M. S. and Lim, T. T., The vortex-shedding process behind two-dimensional bluff bodies, *J. Fluid Mech.,* 116, 77, 1982.

Pineau, G., Mise en évidence et évaluation des effets tridimensionnels dans l'écoulement instationnaire autour d'un cylindre circulaire d'envergure finie, Thesis of the University of Poitiers, France, 1992.

Pineau, G., Texier, A., Coutanceau, M. and Ta Phuoc Loc, Experimental and numerical visualization of the 3-D flow around a short circular cylinder fitted with end-plates, *Flow Visualization VI,* Tanida, Y. and Miyashiro, H., Eds., Springer-Verlag, New York, 1992, 343.

Polidori, G., Etude par visualisation de sillages tridimensionnels. Application à un profil d'aile rectangulaire, Thesis of the University of Poitiers, France, 1994.

Polidori, G. and Coutanceau, M., Formation et échappement tourbillonnaires à l'extrados d'un profil mince en incidence, *12th French Congress of Mechanics,* Burnage, H., Ed., Strasbourg, 1992.

Polidori, G. and Coutanceau, M., Naissance et développement de la tridimensionnalité dans le sillage d'une plaque à 90° limitée par deux flasques rectangulaires, *Visualisation et Traitement d'Images en Mécanique des Fluides,* Schon, J.-P., Ed., Saint-Etienne, 1995.

Polidori, G., Pineau, G., Abed Méraïm, K. and Coutanceau, M., Shedding process of the initial vortices from impulsively started cylinders at Re = 1000 : end and body geometry effects, *Bluff-Body Wakes, Dynamics and Instabilities,* Eckelmann, H., Graham, J. M. R., Huerre, P., and Monkewitz, P. A., Eds., Springer-Verlag, Göttingen, 1992, 285.

Prenel, J.-P. et al., Dynamic and multiple laser sheets: an experimental approach of unstationary and three dimensional flows, *Atlas of Visualization,* 3, Nakayama, Y. and Tanida, Y., Eds., Visualization Society of Japan, 1996, chap. 13.

Prenel, J.-P., Porcar, R., Polidori, G., Texier, A. and Coutanceau, M., Wavelength coding laser tomography for flow visualizations, *Opt. Comm.,* 91 (no. 1–2), 29, 1992.

Tobak, M. and Peake, D. J., Topology of two-dimensional and three-dimensional separated flows, *12th AIAA Fluid and Plasma Dynamics Conf.,* paper 79-1480, Williamsburg, 1979.

Tobak, M. and Peake, D. J., Topology of three-dimensionnal separated flows, *Annu. Rev. Fluid Mech.,* 14, 61, 1982.

chapter four

Visualization of Velocity and Vorticity Fields

Anjaneyulu Krothapalli, Luiz Lourenco, and Chiang Shih

Fluid Mechanics Research Laboratory, Florida A&M University and Florida State University, Tallahassee, Florida

Introduction

Among the currently available experimental techniques, particle image velocimetry (PIV) is the most suited for the measurement of synoptic velocity fields over a two-dimensional region of flow. This technique provides sufficient accuracy and spatial resolution to visualize flows that evolve stochastically both in space and time. Several examples of flow fields will be presented which reveal some important fluid physics. The measured velocity fields serve as basis for the calculation of derived fields such as vorticity fields. Graphical representation of such fields allow us to gain not only quantitative, but also visual insight into the flow field. Example flow fields presented are a shock-cell of an underexpanded axisymmetric jet, the interaction of a free shear layer with a boundary layer within a transitional wall jet, the dynamic stall process generated by a pitching-up airfoil, and the leading edge vortex flow of a delta wing.

The principle behind the PIV technique is that instantaneous fluid velocities can be evaluated by recording the position of images produced by small tracers, suspended in the fluid, at successive time instants. The underlying assumption is that these tracers follow closely and, with minimal lag, the fluid motion. The assumption holds true for a wide variety of flows of interest provided that the tracers are small enough and/or that their density approaches that of the fluid. The technique can operate in two distinct modes. In the auto correlation mode, a recording of two or more instantaneous image patterns created by the seeds is recorded in the same frame. In the cross-correlation mode, the individual instantaneous patterns are kept in separate frames. The time interval between exposures is chosen such that the tracer particles will have moved only a few diameters, far enough to resolve their motion, but less than the smallest fluid macroscale. The distance between the multiple-exposed images from one seeding particle, which is proportional to the local fluid velocity, can then be measured using standard image processing techniques. The detailed description of the PIV technique can be found in a review article by Lourenco et al. (1989). The particular experimental setup used to investigate the different flows discussed here is given in their respective sections.

Supersonic Flow: The Shock-Cell of an Underexpanded Jet

The shock-cell studied in this section was generated by an underexpanded jet issuing from an axisymmetric convergent nozzle at a moderate pressure ratio into quiescent ambient air. This study demonstrates the ability of PIV for the measurement of velocity fields in a complex supersonic flow that contains expansion fans and shock waves. For a more detailed description of the shock-cell structure, reference can be made to a recent paper by Krothapalli et al. (1996).

On-Line Particle Image Velocimetry

A fully digital and operator-interactive "on-line" PIV system developed by Lourenco (1990) was used in this experiment. This instrument uses a high resolution CCD area sensor and state-of-the-art microcomputer hardware and software for image acquisition and processing. Unlike earlier digital approaches, it is capable of recording flows within a wide range of velocities, from a few millimeters per second to several hundreds of meters per second while retaining the spatial resolution of a conventional 35-mm film. The basis for the evaluation of the displacement of particle images is the autocorrelation method (see Lourenco et al., 1994, for details).

The main components of this fully integrated PIV are the laser source, with associated sheet-forming optics, the high resolution video sensor, the microcomputer, and the image acquisition and processing hardware. A video camera (KODAK 1.4 Megaplus) with a CCD array containing 1320(H) × 1035(V) active pixels was used to acquire the multiple exposed images. The pixels are 6.8 µm square and have a center-to-center spacing of 6.8 µm; thus the spatial resolving power of this sensor array is equivalent to 74 line pairs per millimeter. The camera features a built-in 8-bit analog-to-digital converter that produces a digital video output signaling containing 256 gray levels. A customized digital interface and frame buffer board in the host computer receive and store the digital frame data from the camera. Using a calibration experiment, Lourenco and Krothapalli (1995) have demonstrated that the velocity measurement error is bounded and of the order of ±1 to 2% full scale. This technique in combination with a velocity bias device made possible velocity measurements in regions of high shear.

A convergent axisymmetric nozzle with an exit diameter D of 2.25 cm was used to generate an underexpanded jet. The operating conditions were as follows: stagnation pressure, p_o = 0.8 MPa, and temperature, T_o = 288 K. The jet issued into quiescent ambient air (p_a = 0.1 MPa, T_a = 293 K). The exit Reynolds number based on isentropic conditions and the exit diameter were about 2.3×10^6. A top-hat velocity profile with laminar boundary layers was maintained at the nozzle exit. Based on our earlier investigations, the nozzle exit boundary layer momentum thickness was estimated to be about 0.1 mm. The resulting exist Reynolds number Re_θ, based on the momentum thickness, θ, is about 1×10^4.

In applications of PIV to the measurement of high speed flows, the selection and implementation of the proper seeding strategy are major factors contributing to successful measurements. In this study the air jet was seeded with 0.3 µm Al_2O_3 particles. These particles were introduced through an agitated fluidized bed with low mass flow and velocity so as to allow only the smallest particles to become airborne. The suspended particles were passed through a cyclone particle separator to remove any large particles. The details of the seeding apparatus were given by Ross (1993). The ambient air was seeded with smoke generated by a Rosco 1500 Fog/Smoke generator. This generator produces particles in the range of 0.5 to 1.5 µm.

The laser sheet illumination was created by using a Lumonics double-pulse (pulse width = 20 ns) Ruby laser. The width of the light sheet was about 15 cm and its thickness

Chapter four: Visualization of velocity and vorticity fields 71

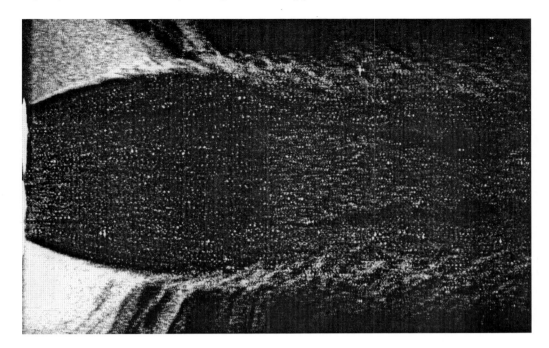

Figure 1 Double-exposed PIV image of an underexpanded jet.

about 1 mm. To capture the velocity of the entrained fluid from the ambient air and within the shear layers, a velocity bias was applied using a high speed (6000 rpm) spinning mirror.

The double-pulse images acquired using a 1.4-MB (1340 × 1035) Kodak digital video camera in conjunction with FFD MkIII "on-line" PIV system were processed on a rectangular Cartesian grid with 50 × 60 points. Using a i486, 66-MHz based computer with an i860 accelerator board, velocity vectors were computed at a rate of 30 vectors/s. High resolution Schlieren images show the wave structure consists of a centered expansion fan at the nozzle lip that reflects from the outer boundary of the jet to form a compression fan that coalesces into a shock, normally referred to as barrel or intercepting shock. Due to overexpansion of the gas beyond that of the corresponding isentropic flow Mach number, a normal shock appears in the center portion of the jet, normally referred to as Mach disk. This shock terminates at a triple point that is formed by Mach disk and two oblique shocks.

A typical double-exposure image of the jet flow field is shown in Figure 1. The different particles used to seed the ambient air (smoke) and the jet (Al_2O_3) are used as a discriminator between the two distinct fluid regions. Measurement within the shear layers is made possible by the use of the velocity bias as well as this kind of approach to seeding. Another by-product of the double-exposed images is the qualitative measure of the fluid density, shown as variations in the particle concentration. For example, a distinct boundary in particle concentration is noticed at the location of the Mach disk. In the subsonic region immediately after the shock, a noticeable increase in the particle concentration is observed suggesting a relatively large value for the density as compared to that upstream of the shock. The slip-line shear layers inside the jet that originate from the triple point are also clearly seen. Because of their relative low convective Mach number large coherent structures are observed ($M_c \approx 0.3$, where $M_c = (\mathbf{U}_1 - \mathbf{U}_2)/(\mathbf{a}_1 + \mathbf{a}_2)$; $\mathbf{U}$, $\mathbf{a}$ represent mean velocity and speed of sound, respectively, and subscripts 1 and 2 designate the high and low speed sides of the shear layer). In order to adjust for the pressure imbalance, the outer shear layer follows a concave curvature. This outer shear layer displays very little organized motion, as this is a typical characteristic of high convective Mach number flows. The average convective Mach number of the shear layer is about 0.8.

The two-dimensional mean velocity field in the central plane of the shock-cell shown in Figure 2(a) was obtained from averaging 60 double-exposure images similar to the one shown in Figure 1. They were processed using a Cartesian grid (x, r, where x denotes the ordinate along the axis aligned with the centerline of the jet with its origin at the nozzle exit and r the jet radius) with 50×60 points. The corresponding physical spacing between points is 1.73×1.15 mm. Figure 2(a) clearly depicts the mean velocity variation within the shock-cell structure; the velocity distribution in outer and inner (slip stream) shear layers was obtained with sufficient spatial resolution for the calculation of the out-of-plane component of the velocity. Due to the finite size of the seed particles, the measured velocity differs from the corresponding fluid velocity downstream of strong shocks. The effect of particle lag was estimated accurately using a modified stokes drag formula and the data agree well with the model.

The normalized out-of-plane component of vorticity ($\omega = \Omega D/U_e$, where Ω is the out-of-plane component of the vorticity and U_e is the mean nozzle exit velocity) was obtained and shown in Figure 2(b) as color-coded isovorticity contours. In addition to the outer shear layers, the inner slip-line shear layers are also evident. It is of interest to note that the vorticity strength of the slip-line shear layer is comparable to that of the outer shear layer. For example, the maximum normalized value of vorticity in the outer shear layer is found to be about 15, while for the slip-line shear layer it is about 11. The core of the jet contains no significant vorticity suggesting that the isentropic assumptions in these regions are quite valid.

Low Speed Air Flow: A Transitional Wall Jet

A plane-wall jet is formed when a stream of fluid is blown tangentially along a plane wall. From past observations, two distinct flow regimes can be identified: an inner region where the flow resembles the conventional boundary layer and an outer region where the flow is similar to that of a free-shear layer. In this section, a typical result showing the double-row vortex structure that is characteristic of a transitional wall jet is presented. The jet is issued from a rectangular channel having a parabolic velocity profile at the exit. The Reynolds number, based on the exit mean velocity and the channel width, is 1450. Instantaneous PIV velocity and vorticity–field measurements provide a basic understanding of the mechanism involved in the formation of the vortices in the inner region of the wall jet and their subsequent interaction with the outer shear-layer vortices. For a detailed description of the flow field, reference can be made to Gogineni (1994).

A rectangular channel 0.5 cm high (h) and 52.7 cm long (L) with a 10.0 cm span (S) is connected to the settling chamber by a two-stage contraction having an overall contraction/area ratio of 80. The aspect ratio (S/h) of the channel is 20. A loud speaker is attached to the settling chamber to create the acoustic disturbances for phase locking the flowfield. A Cartesian coordinate system (x-axis in the streamwise direction, y-axis in the cross-stream direction, and z-axis in the spanwise direction) is chosen to describe the measurements of the velocity field.

The PIV image is obtained by illuminating the seeded flowfield with a thin laser light sheet. A dual-pulsed laser system consisting of two Spectra-Physics DCR-11 Nd:YAG lasers is used to provide the double-illumination pulses. A Bk-7 beam-combining cube along with an adjustable mirror system is used to make the two laser beams collinear. The combined laser beam is directed toward the test section to create the laser sheet in the selected plane using cylindrical lenses. The delay between the two laser pulses is controlled by a pulse generator. The jet is seeded with smoke particles produced by a smoke generator

Chapter four: Visualization of velocity and vorticity fields

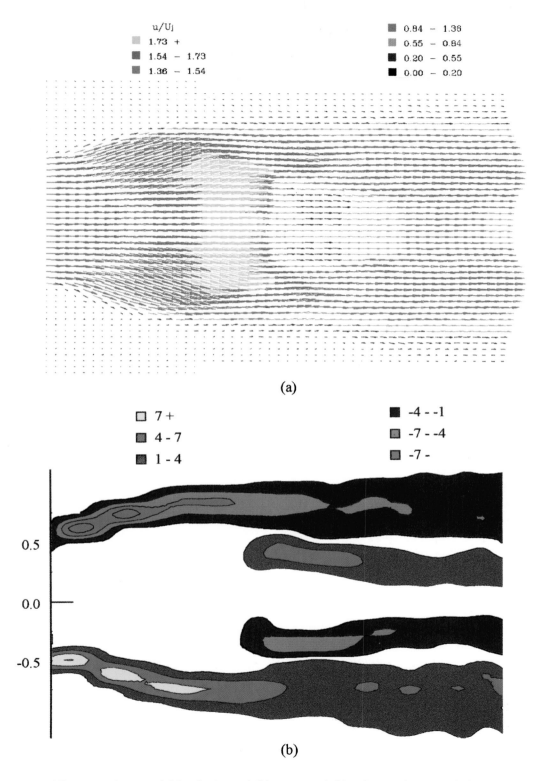

Figure 2 Averaged (a) velocity and (b) vorticity fields of an underexpanded jet.

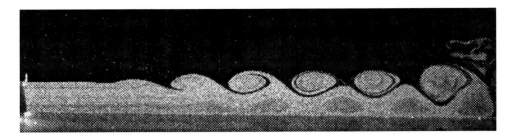

Figure 3 Flow visualization of a transitional wall jet.

(Rosco 1500). The size of the smoke particles is estimated to be on the order of 0.5 μm. Smoke and ambient air are mixed in a large tank before entering the jet supply system. A second smoke generator of the same type is used to seed the outside ambient flow surrounding the jet. A 35-mm camera (Nikon F3) is used to record the flowfield.

The laser sheet flow-visualization picture for an unforced, natural wall jet is shown in Figure 3. This photograph, covering the region from the jet exit to a location 16 heights downstream, clearly shows the emergence of discrete vortices in both the inner and the outer regions of the jet. In the outer region, Kelvin–Helmholtz–type instability waves grow near the nozzle exit and roll into discrete vortical structures further downstream. In the inner region, an array of discrete vortices that align out of phase with respect to the corresponding outer structures can be seen. The outer shear-layer vortices rotate in a counterclockwise direction, and the boundary-layer vortices rotate in a clockwise direction. This organized double-row vortex structure emerges near the nozzle exit and is convected downstream until the structures become indistinguishable further downstream.

Acoustic excitation from a loud speaker is used to provide a clean phase-synchronization to the flow development. Figures 4(a) and (b) show ensemble-averaged PIV velocity and associated vorticity fields; both are phase-averaged over 30 individual PIV data fields, corresponding to a specific excitation phase. These data clearly depict the appearance of two discrete vortices in the outer shear layer. At the channel exit, the shear-layer vorticity coalesces immediately into a vortex. The rapid rate of vortex growth results from the application of excitation at a high amplitude. Under the influence of this vortex, the local boundary layer is thickening, eventually separates from the surface, and forms a discrete vortex rotating opposite to that of the shear-layer vortex. These two counterrotating vortices later convect together, forming a vortex couple.

Based on extensive observations the formation of the boundary-layer vortex can be explained as follows. The rolling up of the free shear-layer vortex causes acceleration of the flow immediately above the wall and results in the emergence of a local low surface pressure. This observation was confirmed by unsteady surface-pressure measurements. Consequently, the boundary-layer flow experiences an adverse pressure gradient ahead of the vortex. The low-momentum fluid near the surface cannot negotiate this abruptly imposed pressure gradient and, as a result, the flow detaches from the wall. As the free shear-layer vortex convects downstream, the favorable gradient following the adverse pressure gradient tends to force the detached flow to reattach, forming a recirculation region. This recirculation region grows rapidly along the transverse direction and pushes the boundary-layer flow away from the wall. The separating boundary layer is, therefore, subjected to the inviscid instability usually experienced by a free shear layer, and coalesces into a vortex in the inner region. The inner vortex interacts strongly with the shear-layer vortex, and this interaction dominates the subsequent development of the wall jet.

Chapter four: Visualization of velocity and vorticity fields 75

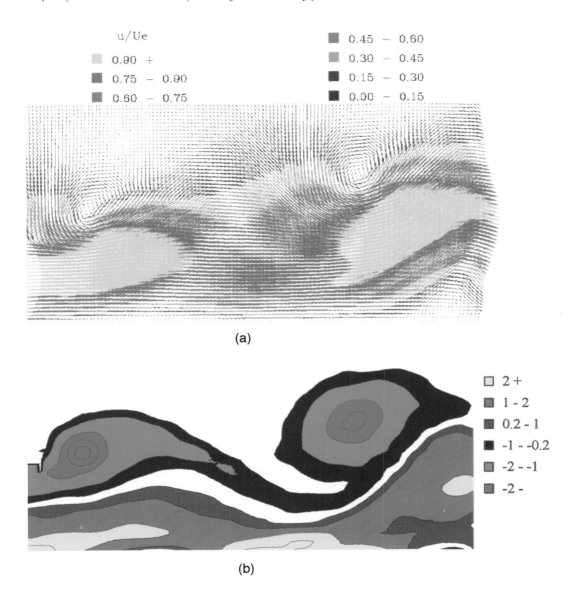

Figure 4 Phase-averaged (a) velocity and (b) vorticity fields of a forced wall jet.

Low Speed Water Flows

The following two experiments are performed in a water towing tank facility with dimensions of 3.4 m in length and 0.6 × 0.5 m in cross section. A computer-interfaced Anorail linear motor system was used to drive the towing carriage. This system allows for fine control of the towing velocity. In the present investigation, the velocity of the towing carriage was varied from 5 to 25 cm/s. Three moving platforms, one for model mounting, one for the rotating mirror system used to create the pulsed laser illumination, and the other for image recording, were synchronized using a gear/belt system. All motions were controlled by a desktop IBM PS/2 model 70 computer. The pulsed laser sheet is created by means of a 24-faceted rotating mirror system. This mirror sweeps an 18-W Argon-ion laser beam into a laser sheet which was projected along the midspan of the airfoil, providing multiple particle image illumination required for the PIV image recording.

Metallic coated particles (TSI model 10087), with an average diameter of 11 μm, were used as the flow tracers. A phase-triggered 35-mm camera was used to record the image at a controllable rate. Synchronization between components was accomplished using a Tektronix modular electronics system. This system also provided the phase reference between the motion of the airfoil and the PIV photographic timing sequence.

Leading Edge Flow of a Pitching-Up Airfoil

With recent interest in the development of highly maneuverable aircraft, much emphasis has been placed on understanding unsteady separated flows that occur over a wing pitching up rapidly from zero to high angle of attack. Such a rapid change in angle of attack will result in an unsteady flow field which is significantly different from that of a wing under a quasi-steady-state motion. The departure is mainly due to the initiation of a local, unsteady boundary layer separation and its subsequent interaction with the external flow. This strong inviscid/viscous interaction eventually leads to the massive boundary layer separation and the formation of large-scale vortical structures, commonly referred to as dynamic stall vortices. These structures dominate the unsteady flow behavior over the wing. Since these vortices induce large suction forces on the upper surface of the wing, they are critical to its aerodynamic performance. A careful management of these energetic structures can result in the development of an effective control scheme to improve the flight envelope.

The dynamic stall process of a pitching-up airfoil as described by Shih et al. (1992, 1995) includes the following main features. It is initiated by the unsteady boundary layer separation near the airfoil's leading edge. During rapid pitching-up motion of the airfoil, vorticity production is greatly enhanced by the favorable pressure gradient at the leading edge. At the same time, vorticity accumulated locally due to slowdown of the downstream convection process because of the presence of the adverse pressure gradient and local boundary layer flow reversal. This accumulation process is eventually interrupted by the sudden emergence of unsteady flow separations, which immediately releases the accumulated vorticity into the outer flow. This erruption of boundary layer vorticity triggers a sequence of spontaneous events such as the local viscous/inviscid boundary layer interaction, the formation and convection of large energetic vortices, and, finally, the "stall". Main parameters that govern the flowfield of an airfoil undergoing pitching-up motions include: the airfield geometry, freestream Reynolds number, and the nondimensional pitch rate. In the present work an NACA 0012 airfoil was used. The Reynolds number, based on the chord of the airfoil, is 5000. Experiments were conducted at dimensionless pitch rates, $\left(\alpha^+ = \dfrac{\alpha C}{U_\infty} \right)$, normalized with the airfoil's chord and the free-stream velocity of 0.131.

Instantaneous velocity and associated out-of-plane vorticity fields were measured using the whole-field PIV technique. Special emphasis has been placed on the study of flow structure near the airfoil's leading edge. A typical PIV image is shown in Figure 5. In order to accurately capture the fine structures near the surface of the airfoil, the PIV data are processed with a very high spatial resolution (a physical spacing of 0.5 mm, or 0.5% chord length, in both directions). From the detailed PIV measurement, the emergence of unsteady separation, up to the formation and ejection of the dynamic stall vortex, is shown in Figures 6(a) to (d). These pictures cover approximately the region from the leading edge to 60% chord downstream.

As the airfoil pitches beyond its static stall angle, a strong leading-edge suction pressure peak is developed by the presence of a rapidly accelerating fluid stream flowing over the nose. Downstream of the nose region up until the airfoil's trailing edge, flow reversal develops near the surface where an increasingly adverse pressure gradient dominates. However, unlike steady flow, the external flow stream continues to follow the

Chapter four: Visualization of velocity and vorticity fields 77

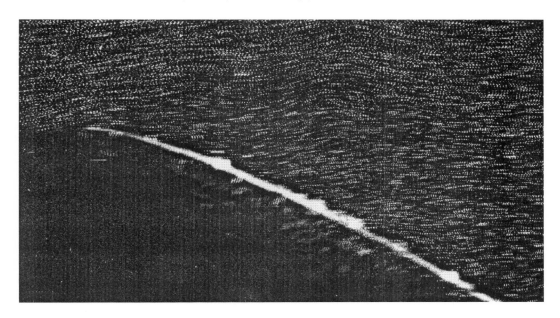

Figure 5 Multiple-exposed PIV image of a pitching-up airfoil.

airfoil's contour and there is no apparent "breakaway" of the boundary layer flow from its surface (Figure 6[a]). Consequently, a local "free" shear layer emerges between the displaced leading edge boundary layer and the reversing fluid layer. This shear layer is subjected to inflectional instability and leads to the generation and growth of individualized vortical structures (Figures 6[a] and [b]). Because of the increasing adverse pressure gradient, the local reversing flow begins to accelerate rapidly inside the leading edge region. This leads to a strong local boundary layer separation and the emergence of the dynamic stall vortex (Figures 6[b] and [c]). Under the influence of the primary vortex, the reversing flow also separates from the surface and develops into a secondary vortical structure (Figures 6[c] and [d]). Strong interaction between the primary vortex and the secondary structure eventually leads to the eruption of the reversed boundary layer into a spike-like structure (Figure 6[d]). This type of strong inviscid/viscous interaction is a generic characteristic of unsteady separated flow.

Leading Edge Vortices of a Delta Wing

The most distinguishing feature of the delta wing flow field is the existence of a pair of well-organized and highly energetic counterrotating leading edge vortical structures. The leading edge vortex flow induces large suction force on the delta wing surface and enhances the overall-performance of the aircraft. Because of its superior performance, the delta wing is the most widely used generic form for the study of flight aerodynamics. Unfortunately, at a high angle of attack, the vortices develop large-scale instability that is characterized by the rapid deceleration and eventual stagnation of the axial velocity along the vortex core, which leads to strong oscillation and total breakdown of the vortical structure. This phenomenon is commonly known as vortex breakdown, or vortex burst. After breakdown, the leading edge vortices lose their effectiveness in generating high lift force. In addition, the ensuing flow unsteadiness associated with the vortex breakdown generates large buffeting forces on the wing, resulting in the deterioration of the quality of aerodynamic control.

In this section, the flow field over a delta wing at a fixed angle of attack of 12.5° is studied. The emphasis is placed on understanding the aerodynamic behavior of the basic

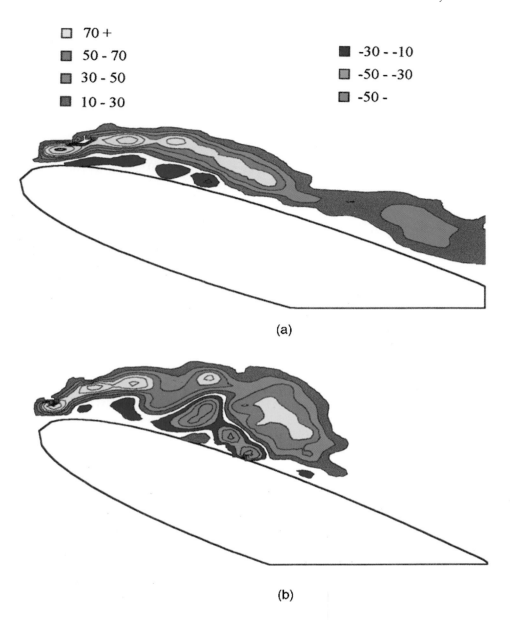

Figure 6 Instantaneous vorticity fields of a pitching-up airfoil. (a) $t^+ = 2.52$, $\alpha = 18.9°$; (b) $t^+ = 2.86$, $\alpha = 21.4°$; (c) $t^+ = 3.36$, $\alpha = 25.2°$; (d) $t^+ = 3.53$, $\alpha = 26.5°$.

flow structure (such as the breakdown of leading edge vortices) of a delta wing (Shih and Ding, 1996). The delta wing model has a leading edge sweep angle of 60° and a root chord of 13.0 cm. The Reynolds number, based on the free stream velocity and the root chord, is 9800. The leading edges are beveled 45° leeward from the top surface.

Figures 7(a) and (b) show the ensemble-averaged (over 20 instantaneous ensembles) PIV velocity and associated isovorticity contour fields at 50% chord location where the vortex is not broken down. Both the ensemble-averaged velocity and the vorticity fields show well-organized vortical structures. The velocity field is strongly asymmetric with respect to the vortex core as it has very high outward velocity component (up to 1.5 U_∞) close to the wing (Figure 7[a]). By comparison, the inward velocity above the vortex core

Chapter four: Visualization of velocity and vorticity fields 79

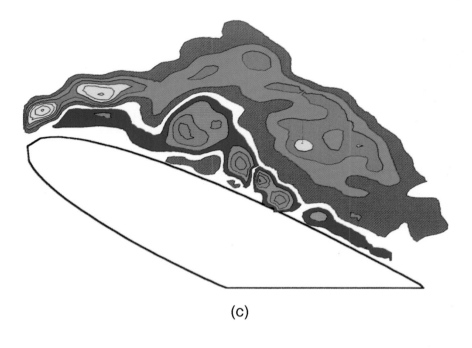

(c)

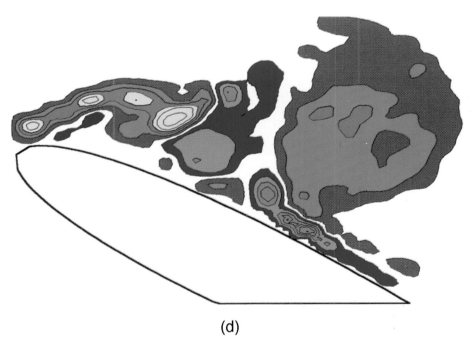

(d)

Figure 6 (continued)

region is much slower and less than half of that value (0.7 U_∞). A strong secondary counterrotating vortical structure is formed as a result of the separation of the outward flowing stream (Figure 7[b]).

Figure 8 presents a sequence of two instantaneous flowfields with distinctive features. It can be seen that the instantaneous vortical structure is drastically different from the organized vortex as characterized by the ensemble-averaged vorticity field (Figure 7[b]).

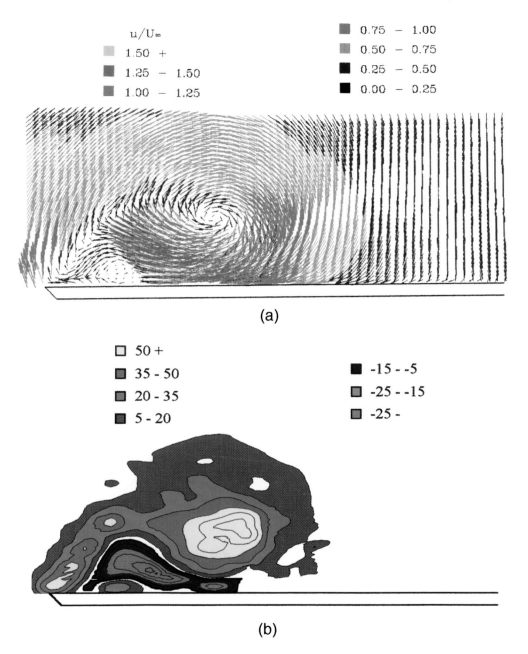

Figure 7 Averaged cross-flow (a) velocity and (B) vorticity fields of a delta wing.

It appears that the secondary vortex actually plays an active role (instead of a "secondary" one as is commonly believed) in the development of the leading edge vortex system by interacting strongly with the primary leading edge vortex.

Conclusions

Using example flows that cover a wide range of velocities and complex features, the ability of PIV to measure accurately the velocity fields was demonstrated. The associated vorticity fields were calculated with accuracy. The measurement of a supersonic flow is demonstrated

Chapter four: Visualization of velocity and vorticity fields

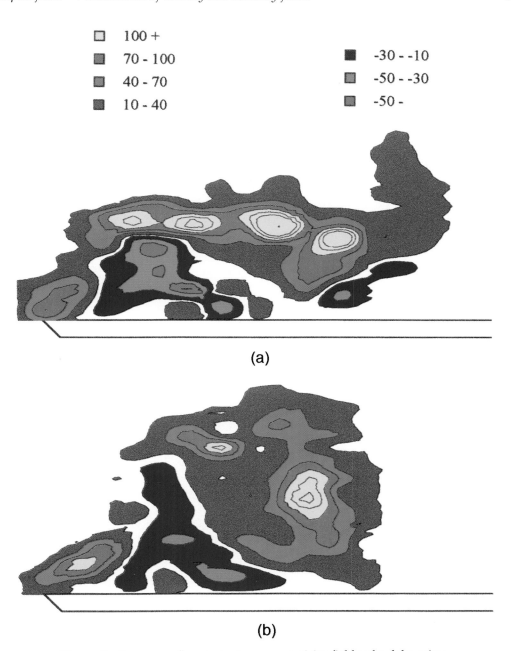

Figure 8 Two cross-flow instantaneous vorticity fields of a delta wing.

through the study of a shock-cell structure generated by an underexpanded jet that contains shocks and expansion fans. The associated vorticity field shows for the first time the structure of the inner slipstream shear layers. An example of a complex low speed air flow was generated by a transitional wall jet that contains an interaction of wall and free shear layers. Such an interaction produces a double-row vortex structure, the details of which were presented.

Complex vortical interactions in unsteady separated flows were presented using low Reynolds number water flows. In the first example, the dynamic stall process over a pitching airfoil was described. The interaction of the dynamic stall vortex with the local boundary layer was characterized using the instantaneous vorticity fields. A similar approach was also used in describing the separation process of a leading edge vortex of a delta wing. The color-coded vorticity fields show the dynamics of the flow with clarity.

Through examples presented in this paper, the whole-field velocity measurement technique, such as PIV, was found to be a valuable tool in the characterization of complex flows that evolve stochastically both in space and time. The range of velocities considered here extend from a few centimeters/second to hundreds of meters/second.

Acknowledgments

The work presented in this paper was supported by AFOSR, NASA Ames Research Center, and NASA Headquarters. Dr. Sivaram Gogineni, Dr. David Wishart, Mr. Chris Ross, Mr. Jim King, and Mr. Zhong Ding have conducted the experiments described in this paper. Color pictures were generated by Mr. Zhong Ding; without his help this paper would have not been possible.

References

Gogineni, S., "Particle Image Velocimetry Study of Two Dimensional Transitional Plane Wall Jet", Ph.D. thesis, Florida State University, Tallahassee, 1994.

Krothapalli, A., King, C.J., Wishart, D.P., and Lourenco, L., "The Shock-Cell Structure of an Underexpanded Jet", *Experiments in Fluids,* submitted, 1996.

Lourenco, L.M., "On-Line Particle Image Velocimetry", *Bull. Am. Phys. Soc.,* Vol. 35, No. 10, 1990, p. 2237.

Lourenco, L.M. and Krothapalli, A., "On the Accuracy of Velocity and Vorticity Measurements with PIV", *Exp. Fluids,* Vol. 18, 1995, pp. 421–428.

Lourenco, L., Krothapalli, A., and Smith, C.A., "Particle Image Velocimetry", Lecture Notes in Engineering: Advances in Fluid Mechanics Measurements, Gad-el-Hak, M., Ed., Springer-Verlag, New York, 1989, pp. 128–199.

Lourenco, L.M., Gogineni, S.P., and LaSalle, R.T., "On-Line PIV: An Integrated Approach", *Appl. Opt.,* Vol. 33, No. 13, 1994, pp. 2465–2470.

Ross, C.B., "Calibration of Particle Image Velocimetry in a Shock-Containing Supersonic Flow", M.S. thesis, Florida State University, Tallahassee, April 1993.

Shih, C. and Ding, Z., "Trailing-Edge Jet Control of Leading-Edge Vortices of a Delta Wing", *AIAA J.,* Vol. 34, No. 7, pp. 1447–1457, 1996.

Shih, C., Lourenco, L., Van Dommelen, L., and Krothapalli, A., "Unsteady Flow Past an Airfoil Pitching at a Constant Rate", *AIAA J.,* Vol. 20, No. 5, May 1992, pp. 1153–1161.

Shih, C., Lourenco, L., Van Dommelen, L., and Krothapalli, A., "Investigation of Flow at Leading and Trailing Edges of a Pitching-up Airfoil", *AIAA J.,* Vol. 33, No. 8, pp. 1369–1376, 1995.

chapter five

Colors in PIV

A. Cenedese and G.P. Romano

Department of Mechanics and Aeronautics, "La Sapienza" University, Rome, Italy

Abstract—*The color of light pulses in particle image velocimetry (PIV) is used to enhance the capabilities of the technique. Image analysis systems employing different laser wavelengths are subdivided into synchronous or asynchronous, consisting, respectively, of simultaneous or consecutive light pulses. The main advantages obtained over the classical "monochromatic" PIV are the increasing velocity dynamical range, the full determination of the direction, the possibility of measuring the velocity component orthogonal to the measurement plane, and the simultaneous detection of particle size and velocity.*

Introduction

Under the name PIV are grouped several measurement methods based on the analysis of images obtained from tracer particles into a fluid. Thereafter, images will be assumed to be acquired by digital video recording systems rather than by classical photographic and optomechanical devices (as in particle (or laser) speckle velocimetry, PSV or LSV): this is the so-called digital particle image velocimetry (DPIV) (Willert and Gharib, 1991; Adrian, 1992). In recent years, DPIV replaced classical methods due to theoretical and practical advantages: simpler experimental setup and data analysis methods, lower overall time consumption, less tedious procedures. Even the higher resolution of photographic films is lost when image processing is performed on conventional frame grabbers. In single-exposed images, although there are no technical bounds to high video rates, economical reasons restrict this rate to $25 \div 50$ Hz, thus limiting single-exposed DPIV to low velocities. The use of multiexposed tracer images can overcome this limitation: while single-exposition methods employ only one tracer particle image in each frame, multiexpositions consist of two (or more) consecutive particle images per frame.

Another problem arises from the density of tracer particles in the fluid under investigations: in many practical situations (especially in air flows) the seeding density is low, i.e., the mean distance between tracer particles is much larger than the particle displacement. Individual particles can be detected and, in principle, "tracked" along their trajectories. This approach is the basis of particle tracking velocimetry (PTV) (Adrian, 1992: Cenedese and Paglialunga, 1990), used in low seeding density, in alternative to PIV based on spectral (or correlation) algorithms, employed in high seeding. PTV enables velocity vectors to be obtained in defined positions, thus determining Lagrangian statistics along the particle trajectory and spatial velocity gradients (Sato and Yamamoto, 1987; Cenedese and Romano, 1991; Cenedese et al., 1993).

However, some disadvantages of multiexposed digital PTV must be pointed out (Grant and Lin, 1990; Sata and Kasagi, 1992):

- Ambiguity in the determination of the direction of the velocity
- Errors in the correct "tracking" of crossing trajectories
- Need for at least three consecutive pulses to decrease the arbitrariness in validation criteria
- Low dynamical range (maximum to minimum velocity ratio)

The last point deals with the small possibility of detecting strong changes of the velocity in the measurement section both in modulus and direction. This limitation is particularly severe in flows with high turbulence intensities and/or large spatial velocity gradients where the most interesting flow structures are smoothed or even lost. Solutions to the above mentioned problems are sometimes possible, for example flow direction can be obtained with complicated optical set up and elaborated identification software as in Grant and Lin (1990). The identification procedure can be improved by using new methods based on neural network algorithms as in Murata et al. (1991) or in Cenedese et al. (1992); however, the dynamical and angular ranges cannot be increased too much.

A solution to the described problems can be found by distinguishing each particle image on the multiexposed frame by means of additional information. This can be done by selective illumination of the tracers with light sheets of different colors.

The use of colors in DPIV is limited to a few applications in recent years. Attempts in resolving the flow field by using two different colors to mark spots were done by Smallwood (1992); however, they failed when used in the complex flow field. A mixed streakline-colored images technique was used by Karasudani et al. (1989); no detection criterion was applied and therefore only qualitative solutions for flow visualization purposed were obtained. Cenedese and Paglialunga (1989) overlapped light sheets of different colors to deduce information on the velocity component orthogonal to the measurement plane (although with a lower resolution than the on-plane velocity components). Recently, Stefanini et al. (1993) employed three pulses of different colors to determine the flow field of an impingint jet; however, they obtained poor quality images and only 18 velocity vectors on a 2-cm^2 region.

In the present paper several applications to DPIV using light sheets of different colors will be discussed. Despite the higher complexity of the illuminating system, it will be shown that the information from the color of the tracer particle might give new insight into velocity and size measurements of tracer particles. In the next sections the experimental setup to overlap light sheets of different colors simultaneously (synchronous color DPIV) and consecutively (asynchronous color DPIV) and applications to measurements in fluids will be described.

Synchronous Color DPIV

In this configuration, two (or more) parallel light sheets of different colors are partially overlapped to obtain a region in which all colors are simultaneously present. The situation is summarized in Figure 1 for the use of two colors: light is pulsed by using pulsed light sources or by means of electromechanical devices. Using this experimental setup, it is possible to detect displacements in all three spatial directions. Along the plane of the light sheets (x,y), conventional DPIV is performed. In the direction orthogonal to the light sheets (z), image analysis based on color is carried out. To this aim, images must be subdivided into different buffers (one for each color) and then the relative mean intensity between colors of each tracer image must be computed. If the relative light intensity distribution

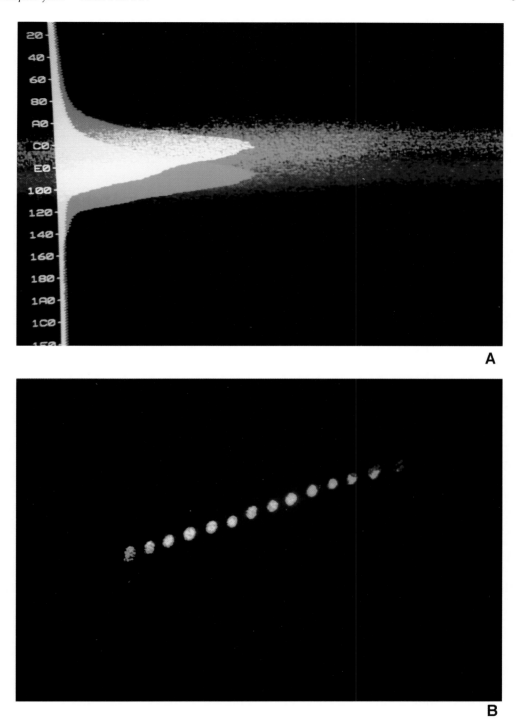

Figure 1 The overlapping of two light sheets of different colors (A) and the relative particle trajectory and illumination during crossing (B).

between the two colors in the direction orthogonal to the light plane is known, it is possible to recover the position of the particle in the z direction.

In comparison with "monochromatic" DPIV, synchronous color DPIV enables one:

- To obtain the position of the particle in the direction orthogonal to the light sheet and, therefore, to determine the velocity component orthogonal to the light sheet
- To determine the size of the tracer particle

However, theoretical and practical limitations are present:

- Light scattering laws (derived from Mie optics) strongly depend on the ratio between the particle size (d) and the light wavelength (λ).
- The relative intensity distribution between the two colors, the distance between the light sheets of different colors, and the receiving camera angular and radial positions must be *a priori* known.

The previous limitations require a preliminary calibration of the optical system for size measurements. This calibration is described in the following; velocity and size measurements are outlined separately.

Third Velocity Component Measurements

Assuming the light intensity distribution across the light sheet (z) to be Gaussian for both colors and the illumination on the plane (x,y) to be constant over the test region, it is (Cenedese and Paglialunga, 1989)

$$I_{1,2} = I_{01,2} f_{1,2}(x,y) e^{\left[-(z \mp a)^2 / \sigma_{1,2}^2\right]} \tag{1}$$

where the subscripts 1 and 2 refer to color 1 and 2 (respectively using − or + in the exponential), I is the intensity distribution across the light sheet, I_0 is the intensity at the peak, f is the spatial attenuation function (depending on the wavelength and on the optical path), a is the half-distance between the peaks of the two light sheets, and σ is the standard deviation of the intensity distribution. The light intensity scattered by each particle crossing the light sheets, Φ, is given by this illumination intensity times a scattering function g: the function g depends on light wavelength, particle size, refractive index between particle and fluid, scattering angles, and distance (Van de Hulst, 1957; Durst et al., 1976; Cenedese et al., 1989). If the two light distributions have the same variance ($\sigma_1 = \sigma_2 = \sigma$), the relative intensity between color 1 and 2 is given by:

$$R = \frac{\Phi_1}{\Phi_2} = \frac{I_{01} f_1 g_1}{I_{02} f_2 g_2} e^{(-4za/\sigma^2)} \tag{2}$$

and if small displacements in comparison with the distance of the observer are considered (such that the functions f and g could be considered identical for the two colors), the ratio between R at point z and R at point $z' = z + \Delta z$ after a time interval between two successive pulses, Δt, is given by:

$$R/R' = e^{4a\Delta z/\sigma^2} \tag{3}$$

By measuring changes in the intensity of each color particle image (averaged over the particle area) during successive expositions, it is possible to measure the displacement Δz orthogonal to the light plane and therefore the velocity component knowing the time interval Δt. There are practical problems when applying this procedure to a measurement

Chapter five: Colors in PIV

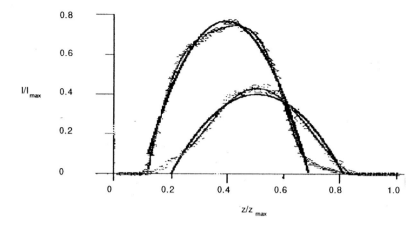

Figure 2 Intensity distribution of green (circles) and blue (triangles) colors. Second- and sixth-order polynomial fittings are plotted; intensity is dimensionless by maximum possible intensity (level 256).

region extending over a distance equal to that of the receiving camera: the change in the intensity of each color could be given not only to displacements in direction z, but also to the angular shape of scattering functions f and g. In Figure 2 the experimental intensity distributions obtained by two light sheets from the green ($\lambda = 514.5$ nm) and blue-violet ($\lambda = 476$ nm) lines of an Ar-ion laser are shown. The distance between the sheets is $2a = 0.2$ mm. A color video camera (Sony DXC) acquires the images on three different CCD centered on blue ($\lambda = 479$ nm), green ($\lambda = 536$ nm), and red ($\lambda = 610$ nm). Image processing starts with transfer to PC through a frame grabber (Matrox MVP) on three different buffers, each with a resolution of 512 × 512 pixels and 256 intensity levels.

In Figure 3 the intensity changes on each color during the crossing of the light sheet are shown. The particle position across the light sheets is observed as a different ratio between green and blue intensity.

However, to compare the results from the measurement of the third velocity component with the other, two error analysis is required. The characteristic size of the acquired region will be indicated as L, while the characteristic particle size will be d. In classical PTV the maximum particle displacement is equal to some fraction of L (say, 3, being necessary at least 3 expositions for validation), whereas the minimum is in the order of the particle size d. It is possible to obtain a rough estimate of the following quantities related to one

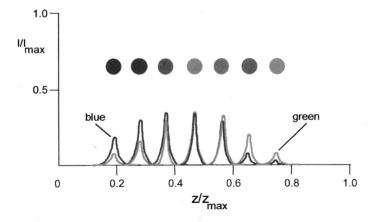

Figure 3 Light intensity variations of green and blue during crossing; intensity is dimensionless by maximum possible intensity (level 256).

velocity component on the (x,y) plane: the dynamical range (D), the resolution (ρ), and the relative error (ε):

- $D = V_{max}/V_{min} \approx L/3d$
- $\rho = 1/V_{min} \approx \Delta t/d$
- $\varepsilon = \Delta V/V \approx d/L$

assuming that $L \gg d$. A typical value is $L/d \approx 50$, giving $D \approx 15$ and $\varepsilon \approx 2.0\%$ in substantial agreement with more detailed computations (Agui and Jimenez, 1987).

Using color to detect the third velocity component, maximum displacement remains the same, whereas the minimum is increased to low particle size (say, $5d$). Moreover, the relative error on the velocity component along direction z must also take into account errors on color determination (that is found to be about 1/10 of the total number of pixels for a particle image, $\Delta C/C \approx 10\%$). Therefore, it is possible to obtain:

- $D = V_{max}/V_{min} \approx L/15d$
- $\rho = 1/V_{min} \approx \Delta t/5d$
- $\varepsilon = \Delta V/V + \Delta C/C \approx d/L + 10\%$

For $L/d \approx 50$, $D \approx 3$ and $\varepsilon \approx 12\%$ are obtained. Therefore, the dynamical range and the resolution for measurements of the component orthogonal to the plane of the light sheets are about five times lower, and the relative error is more than five times larger than those obtained for the components on the plane.

Size Measurements

The primary limitations for particle size measurements using DPIV rely on the nontrivial relation between the real particle and its image appearance. The scattered light intensity of a particle image could depend not only on the size, but also on the amount of incident light. The knowledge of the particle position inside the light sheet brings about the separation of these two contributions and hence the determination of the particle size. As shown in the previous section the particle position in the direction orthogonal to the light sheet can be determined by overlapping two laser sheets of different colors. The starting point is Equation 2. Interest is focalized onto the dependence of function g on the particle size; this is quite complicated and related to Riccati-Bessel and Legendre functions (Van de Hulst, 1957; Durst et al., 1976). However, for practical PIV applications the ratio between particle size and light wavelength is high enough (≥ 100) to give a simple power law dependence of the scattering function g on the particle size (d^n). In theory, $n = 2$, but in practice the exponent could depend on the light wavelength. Moreover, under the assumption of small displacements into the light sheet in comparison with the distance of the observer, the dependence on z (orthogonal to light plane) is retained only into the exponential. The scattered light intensity, Φ, is given by:

$$\Phi_{1,2} = c_{1,2} d^{n_{1,2}} e^{\left[-(z \mp a)^2/\sigma_{1,2}^2\right]} \tag{4}$$

where quantity c retains all the dependences of g and I except for z and d. For small particle displacements, the functions f and g (and hence c and n) are almost identical for the two colors. Therefore, the nondimensional (divided by I_0) relative intensity R, between scattered lights from color 1 and 2, is a function only of the position in the direction z as observed in the present section (Equation 2).

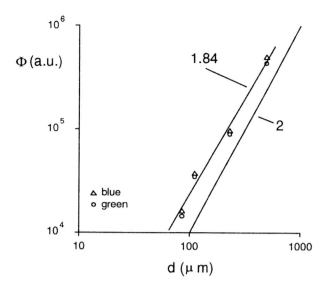

Figure 4 Calibration curves for size measurements using data obtained by blue (triangles) and green (circles). Fit to data (1.84) and square laws.

The procedure for calculating particle size is as follows:

- The position of the particle in the light sheet (direction orthogonal to the plane) is determined through the ratio between color 1 and 2 (Equation 3).
- The particle size is obtained from Equation 4 for color 1 or 2 (or both) if the quantities c and n are known.

The last point requires a preliminary calibration on particles with different sizes to obtain c and n. This is a severe limitation for particle size measurements in flows with tracers with diameters distributed over a wide band. However, the method allows different nonoverlapped size groups to be clearly separated.

In Figure 4 the calibration curve is shown for green and blue colors using the experimental setup described in the previous section; the particle size is measured using a calibrated microscope. The fit to experimental data, from both green and blue measurements, has a slope equal to 1.84; this slope is very close to the square of the particle size. Therefore the approximations used to obtain these curves are reasonable.

The calibration curves have been used to determine the size of glass spheres of size ranging from 85 to 500 µm, the size being measured using the calibrated microscope. The results are summarized in Table 1 for three different experiments for each particle and separately for the blue and green colors. The relative errors are always less then 14% and the mean error is equal to 6.5%. These values are not much greater than errors from size measurements using single point measurements like PDA (around 5%). Slight differences are observed between errors obtained using blue (6.3%) or green (6.5%). The error does not show any dependence on the particle size. The method has also been tested with particles from the same distribution: ten glass particles with known mean distribution centered on 230 µm with a standard deviation of 20 µm.

Results from both green and blue colors give a mean value of 223 µm with a standard deviation of 24 µm; this is in good agreement with the given distribution. The mean value from measurements using green is 221 µm, while that using blue is 225 µm. The difference is less than 2%. Moreover, each single measurement does not differ more than 15% from

Table 1 Particle Size Measurements from Measurements on a Graduate Microscope and from Green and Blue Calibration Curves of Figure 4

	Direct (μm)	Exp. (μm)	Error (%)
Green	85	78	8.2
	85	83	2.3
	85	86	1.2
	110	122	10.9
	110	125	13.6
	110	123	11.8
	230	217	5.6
	230	217	5.6
	230	211	8.3
	500	517	3.4
	500	514	2.8
	500	521	4.2
Blue	85	79	7.0
	85	85	0.0
	85	87	2.4
	110	124	12.7
	110	124	12.7
	110	121	10.0
	230	220	4.3
	230	220	4.3
	230	214	6.9
	500	526	5.2
	500	524	4.8
	500	528	5.6

the mean value. These results confirm that the method is useful for measuring particle size using PIV.

Asynchronous Color DPIV

In this second configuration, two (or more) parallel light sheets of different colors are overlapped, then, in contrast with synchronous DPIV, the sheets are positioned exactly on the same plane. They are separated only in time: pulses of different colors are not simultaneous but delayed one over each other in a time sequence. The situation is clarified in Figure 5 using three colors. This image is obtained using a simulation of a PIV system in a vortex flow. This experimental setup allows the time sequence of the tracer images to be discriminated. Therefore, a conventional DPIV can be applied, provided that the information on the color is taken into account. This technique enables the direction of the velocity and the detection of very small displacements to be obtained.

The first point is simply obtained by separating and analyzing the information on different buffers (one for each color). The information on velocity direction is obtained without any additional shifting device that could complicate the analysis of acquired images. The second point is connected to the possibility of detecting, in principle, also fully overlapped particle images, these being separated by color. As a consequence the dynamical range of the velocity is enhanced due to the lower measurable minimum velocity.

The fundamental drawbacks of this technique are connected to:

- The need for a careful optical alignment

Chapter five: Colors in PIV 91

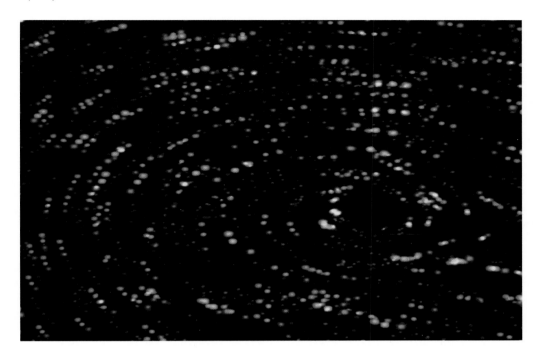

Figure 5 The overlapping of three light sheets of different colors using asynchronous DPIV. Image at tracer particles of different colors are shown.

- The effective color separation in the transmitting optical system tuned with the color separation of the receiving device (video camera)

In the following the asynchronous color DPIV will be employed using a slightly modified version to that previously described. The method consists of overlapping three sheets of different colors, from the same or from different lasers, to feature tracer displacements in a well-recognizable pattern (Busignani et al., 1993; Cenedese and Romano, 1994). Such a pattern is composed of an initial dot of a first color (for example, green [G]), a continuous line of a second color (for example, red [R]), and a final dot of a third color (for example, blue [B]) as shown in Figure 6. The idea is to use both small (spots) and long (streaks) exposures of fluid tracers to obtain their velocity; as already mentioned spots are widely employed in analyzing flow field, while the use of streaks is limited by the practical problem of recognizing their initial and final points (Dimotakis et al. 1981). By connecting two spots with a streak, the method allows:

- To determine the velocity direction (by looking at the sequence [green spot — red streak — blue spot] or to the reverse)
- To detect configurations with crossings (for example, the "four spot configuration" in Figure 6)
- To increase the dynamical range (at the lower boundary because partially overlapped B and G spots are detected and at the higher because displacements as large as the size of the whole image are measured)

Such improvements are connected to an enhanced possibility of recognizing tracer displacements, compared with classical PTV, at similar particle density; this enhancement depends on the connecting red streaks. In Figure 6 is shown how ambiguities are removed: erroneous displacements 1–2, 1–3, 2–4, 3–4 are avoided and the direction is fully solved.

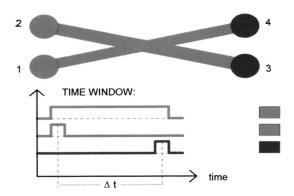

Figure 6 Pattern of two crossing displacements. The ambiguity between spots 1–2, 1–3, 1–4 is solved. The direction is determined.

The flow in the wake of a car model into a water channel is selected as a test case. Flow reversal and domains with large velocity gradients are suitable for comparing color and classical methods.

Velocity measurements on a water channel (length = 120 cm [after contraction and honeycombs], height = 15 cm, width = 10 cm) are performed. A car model (1:43) is positioned on the wall of the channel at about 100 cm from the inlet; the width of the model is about 3.5 cm. The streamwise direction is taken as x axis, while the vertical and horizontal ones are taken as axes y and z, respectively. The Reynolds number is about 100,000 using the flow mean velocity (about 80 cm/s) and channel width; taking the width of the car model as characteristic length, the Reynolds number is about 30,000. Air bubbles (mean diameter between 10 and 50 μm) are used as seeders. The measurement section consists of a portion of the (x,y) plane in the near wake of the model; it extends about 1.5 car widths downstream (x) and 2 car widths in the vertical direction (y). Velocity measurements are nondimensional by the maximum velocity into the test section and distances by the width of the car model.

To take velocity measurements with both classical and the color PTV the same section is illuminated by two different optical systems. The setup of the classical PTV consists of a laser sheet from an Ar-ion laser (green line at 514.5 nm) obtained using spherical and cylindrical lenses. The light is modulated by means of a mechanical shutter (a rotating disk) equipped with 16 small holes; the rotation velocity of the shutter changes from 200 to 2000 rpm. Images of the flow field are taken by a photocamera. Such images are positioned under a microscope connected to a CCD camera; they are acquired by the frame grabber (512 × 512 pixels with 256 gray levels) on a single buffer for further elaboration. Displacements are obtained by analyzing successive tracer positions and areas using a suitable validation procedure (Cenedese and Paglialunga, 1990).

To take measurements with the color PTV, three laser beams are used: green (514.5 nm) and blue (476.5 nm) lines from an Ar-ion laser and red (632.8 nm) from a He–Ne laser (Figure 7). The light is pulsed by means of a mechanical shutter. A special design is employed to obtain the configuration of Figure 6; the red beam crosses the shutter on a continuous hole (adjustable from 45 to 180°), while the blue and green beams on two smaller holes corresponding to the initial and final part of the red one. The beams are then focused onto a cylindrical lens to obtain three light sheets on the same plane. During two exposures the shutter allows at first the green sheet and then the blue one to illuminate the test section, while during the whole time interval the red sheet is on; two tracer images connected by a line will result (Figure 6). A very careful optical alignment is required.

Chapter five: Colors in PIV   93

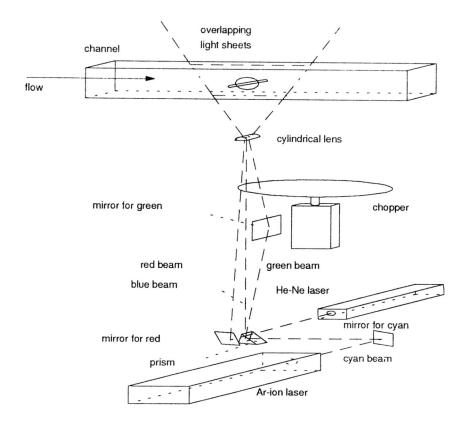

Figure 7 Experimental and optical setup.

Images of the flow are taken and digitized on a 512 × 512 format. Each color is acquired on a different buffer. The sensitivity of the receiver is selected to avoid, as much as possible, the cross-talking between colors; however, on each buffer, small intensity levels of the two other color are present. Such images are processed to obtain tracer displacements in a fully automatic way (Busingnani et al., 1993; Cenedese and Romano, 1994). A black and white image with different light intensities will result. For example, if the red color is set to an intensity value equal to 50, the blue to 70, and the green to 90, pixels with overlapping of red and green colors are characterized by an intensity equal to 140, while pixels with three colors by the value 210. In such a way overlapped spots are detected and the dynamical range is increased, too.

The next step is the identification of connections between spots and the measurement of the displacements. To this end a column-by-column scan is performed looking for R pixels; the contour of the R domain is followed until B and G spots are detected. The positions of the barycenter of such spots are recorded. The identification procedure holds until the whole image is scanned. About 10s are required to perform such operations on a 512 × 512-pixel frame-grabber with 256 gray levels.

Images of the flow field are recorded using the two previously described setups. They are obtained on a vertical section that is a top view of the car model wake; the test section size is about (10 × 7) cm². In Figures 8 and 9 the flow fields using classical and color PTV are shown. It must be noted that the described method does not require either *a priori* knowledge of the field nor information from nearby tracer displacements. Therefore, the identification procedure is the same for the whole image. On the other side, in classical

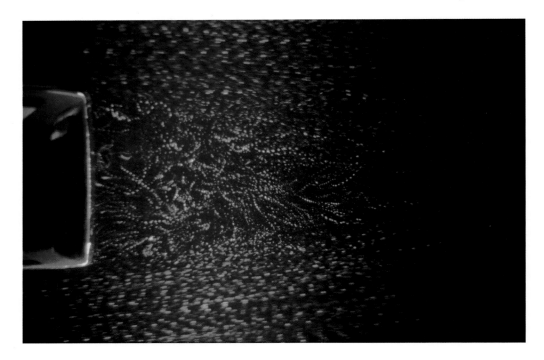

Figure 8 Multiexposed PTV image using classical method. Flow from left to right.

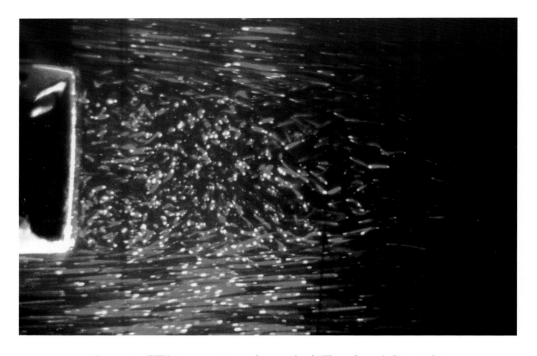

Figure 9 PTV image using color method. Flow from left to right.

PTV, successive positions of the same particle are tracked; distances and angular positions are compared to validate a displacement. Such parameters differ depending on the flow domain to be investigated (for example, an almost straight or a recirculating flow).

Chapter five: Colors in PIV 95

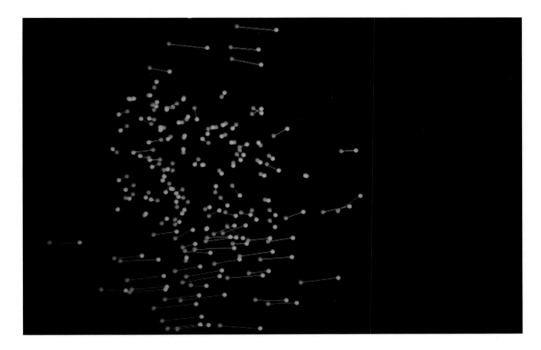

Figure 10 Displacements from PTV color image in Figure 9.

Tracer displacements are obtained with the color method, about 130 velocity data for each image. The maximum measured velocity is equal to about 60 cm/s, the minimum is 0.3 cm/s; therefore, the dynamical range is equal to 200. The angular range is about 180°; flow reversals are fully solved. Displacements obtained using color PTV are shown in Figure 10.

The tracer displacements computed using the classical method depend on the choice of the validation parameters. They give a number of detected displacements less than before: from 50 to 80 (depending on the parameter choice). The dynamical range is between 30 and 100 and the angular range is between 75 and 85° (no velocity direction device is used).

From the tracer displacements, the two velocity components on the measurement section are derived. They are computed on a regular grid by using an inverse distance-weighted algorithm; the grid size is almost equal to the measuring section divided by the number of detected velocities. Using such procedure the interpolation relative error is limited to 5% (Agui and Jimenez, 1987; Spedding and Rignot, 1993).

Velocity vectors on a regular grid are shown in Figure 11 together with vorticity contours: the car model rear part is positioned at $x/d = 0$ (d is the model width) and for y/d ranging from –0.5 to 0.5. The flow field is adequately described: an almost straight flow in the outer wake (y/d larger than 0.6 in modulus), and a recirculation, with also negative velocities, in the near wake are observed. Positive vertical velocity values are observed toward the car model bisecting line. In the near wake (x/d less than 1) an outflow is observed, whereas for larger values of x/d regions of inflow and outflow alternate. In the near wake (x/d less than 1, y/d less than 0.2) the values of the vertical velocity component have the same magnitude of the horizontal.

The vorticity component in the direction z is computed using a finite differences-centered algorithm. Two positive peaks are clearly distinguished; they are connected to the region in which the higher velocity gradients are present, that is, to the boundary between

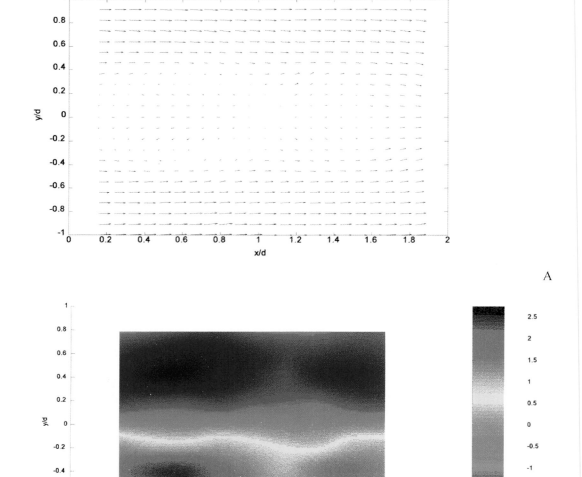

Figure 11 Velocity vectors on a regular grid (A) and vorticity map (B). The car rear part is positioned at x/d = 0 (d is the car model width); the axial mean flow is from left to right. The value y/d = 0 identifies the center line. Vorticity values are nondimensional with u/d (u is the mean axial velocity).

wake and straight flows. As observed by other authors, such flow field leads to separate vortical structures rather than to distributed vorticity.

Concluding Remarks

In the present paper the use of color in DPIV is presented. Two basic configurations have been described: synchronous and asynchronous color DPIV. They allow us:

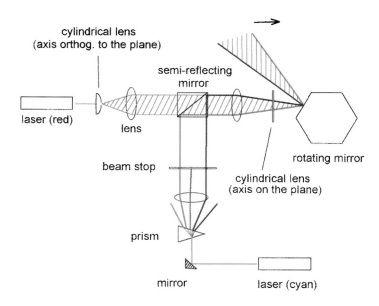

Figure 12 The proposed setup for 3D PTV streak–spot measurements.

- To obtain information on displacements orthogonal to the measurement plane
- To determine the velocity component orthogonal to this plane (with a resolution and dynamical range five times lower and a relative error more than five times greater than the other two components)
- To determine the particle size (with a relative error lower than 15%)
- To enhance the dynamical and angular range of the velocity (at least by a factor 3 for the dynamical range and a factor 2 for the angles, that is, flow reverse is resolved)
- To determine the velocity direction

On the other hand, some drawbacks have to be pointed out:

- Need for a careful optical alignment
- Effective color separation (RGB separation, tuning of transmitting, receiving, and acquisition devices)
- Calibration for size measurements (limiting the usefulness to separation of different distributions)
- The poor resolution and greater errors on the component orthogonal to the measurement plane

Finally, the setup to extend the streak–spot method to 3D measurements is presented in Figure 12.

References

Adrian R.J., 1992, "Particle-imaging techniques for experimental fluid mechanics", *Ann. Rev. Fluid Mech.*, Vol. 23, p. 261.

Agui J.C. and Jimenez J., 1987, "On the performance of particle tracking", *J. Fluid Mech.*, Vol. 185, pp. 447–468.

Busignani S., Cenedese A., Querzoli G. and Romano G.P., 1993, "Combined streak-spot image analysis method using multicolor light source", Proc. Winter Annual Meeting of ASME, New Orleans, pp. 143–148.

Cenedese A., Cioffi F. and Romano G.P., 1989, "Doppler signal predictions using the Lorenz-Mie theory for applications to measurements in two-phase flows", *Arch. Mech.,* Vol. 41, 6, pp. 821–835.

Cenedese A. and Paglialunga A., 1989, "A new technique for the determination of the third velocity component with PIV", *Exp. Fluids,* Vol. 8, pp. 228–230.

Cenedese A. and Paglialunga A., 1990, "Direct digital analysis of a multiexposed photograph in PIV", *Exp. Fluids,* Vol. 8, pp. 273–280.

Cenedese A. and Romano G.P., 1991, "Space derivatives and lagrangian correlations evaluation with PIV technique", *Proc. 4th Conf. on Laser Anemometry,* Cleveland, Alexander Dybbs and Bahman Gorashi ed., New York, p. 341.

Cenedese A., Paglialunga A., Romano G.P., and Terlizzi M., 1992, "Neural net for trajectories recognition in a flow", *Proc. 6th Int. Symp. on Application of Laser Techniques to Fluid Mechanics,* Lisbon, chap. 27.1.

Cenedese A., De Gregorio F. and Romano G.P., 1993, "Vorticity measurements in a turbulent boundary layer using PIV", *Proc. 5th Int. Symp. on Refined Flow Modeling and Turbulence Measurement,* Paris, pp. 301–308.

Cenedese A. and Romano G.P., 1994, "Comparison between classsical and three-color PIV in a wake flow", *J. Flow Visualization Image Process.,* Vol. 1, pp. 371–384.

Dimotakis P.E., Debussy F.D. and Koochesfahani M.M., 1981, "Particle streak velocity field measurements in a two dimensional mixing layer", *Phys. Fluids,* Vol. 24, pp. 995–999.

Durst F., Melling A. and Whitelaw J.H., 1976, *Principles and Practice of Laser Doppler Anemometry,* Academic Press, London, pp. 57–95.

Grant I. and Lin A., 1990, "Directional ambiguity resolution in particle tracking velocimetry", *Exp. Fluids,* Vol. 10, pp. 71–76.

Karasudani T., Funakoshi M. and Oikawa M., 1989, "Breakdown and arrangement of vortex streets in a far wake", *Repr. J. Phys. Soc. Jpn.,* Vol. 58–5, p. 1597.

Murata S., Kise H. and Miyake H., 1991, "Automatic method for determining flow directions by using a neural network model", *Proc. FLUCOME 91,* pp. 657–662.

Sata Y. and Kasagi N., 1992, "Improvement towards high resulution measurement in three-dimensional particle tracking velocimetry", *Proc. Int. Symp. on Flow Vizualization 6,* Yokohama, ed., Springer-Verlag, Berlin, pp. 792–796.

Sato Y. and Yamamoto K., 1987, "Lagrangian measurement of fluid-particle motion in an isotropic turbulent field", *J. Fluid Mech.,* Vol. 175, pp. 183–199.

Smallwood G.J., 1992, "A technique for two-colour particle image velocimetry", Master degree thesis at Department of Mechanical Engineering, University of Ontario, Canada.

Spedding G.R. and Rignot E.J.M., 1993, "Performance analysis and application of grid interpolation techniques for fluid flows", *Exp. Fluids,* Vol. 15, pp. 417–430.

Stefanini J., Cognet G., Vila J.C., Merite B. and Brenier Y., 1993, "A colored method for PIV technique", *Flow Visualization and Image Analysis,* F.T.M. Nieuwstadt ed., Kluwer Academic Publishers, the Netherlands, pp. 247–258.

Van de Hulst C.M., 1957, *Light Scattering by Small Particles,* John Wiley & Sons, New York.

Willert C.E. and Gharib M., 1991, "Digital particle image velocimetry", *Exp. Fluids,* Vol. 10, pp. 181–193.

chapter six

Flow Visualization in Turbulent Large-Scale Structure Research

Lorenz W. Sigurdson

Mechanical Engineering Department, University of Alberta, Edmonton, AB, Canada

> *Abstract*—*Examples are given of how flow visualization has been used to discover the large-scale coherent structures in four different turbulent flow types. These are the separated and reattaching flows over a blunt-faced circular cylinder aligned coaxially with the free stream and its two-dimensional analog the blunt flat plate; the vortex ring resulting from a nuclear explosion; a water drop impacting on a pool of water; and a bursting air bubble initially resting on a free surface. A brief review is given of the general methodology used for interpretation of data obtained by direct injection of tracer materials. Each flow is discussed in terms of the visualization technique used and what conclusions about the structure and other flow aspects could be deduced. The topological similarity between the three-dimensional and unsteady vortex structures of these four flow types is stressed.*

Introduction

Vorticity, a local "spinning" of a fluid, is present in most fluid flows. The more well-known cases of vorticity creation are as follows. It can be created at the solid physical boundaries of a fluid where a pressure gradient is present, for example, the hood of an automobile. It is created within the fluid where there are components of gradients in density perpendicular to the local pressure gradient, for example, a buoyant thermal which keeps a sailplane aloft. Fluid–fluid interfaces can also give birth to vorticity. Where the vorticity goes from there influences many things of engineering importance such as: lift and drag on an aircraft wing, wind forces on buildings, noise production from turbulence, and valve closure within the human heart.

Research suggests that vorticity does not disperse completely randomly but often forms organized large-scale structures, as discussed by Roshko (1976). Within the apparent chaos, an orderly identifiable large-scale structure often exists. These structures are essentially regions of vorticity. Saffman (1980) stated that it is now proposed that turbulence be modeled as the creation, evolution, interaction, and decay of these structures. In his book (Saffman, 1992), he continues:

> The discovery of coherent structures in turbulence has fostered the hope that the study of vortices will lead to models and an understanding of turbulent flow, thereby solving or at least making less mysterious one of the great unsolved problems of classical physics.

Knowledge of the flow structure offers new potential to control turbulent flows to the engineer's advantage. This could be done using steady or unsteady and/or spatially varying forcing techniques such as sound or moving control surfaces. This could result in drag reduction on automobiles and aircraft, improved combustion efficiency, and improved efficiency of an array of mechanical devices.

It is supposed that each particular flow geometry will have a characteristic vortical structure. Its determination is not unlike the task in chemistry of determining the structure of a molecule once the constituent elements are known. Much can be surmised with a few bits of knowledge about the expected behavior of the flow, and an understanding of what the location of the vorticity means to things like conservation of momentum and the velocity field (from the Biot-Savart law). Neglecting viscous diffusion, the vorticity moves as a material element with the flow; therefore, many simple vortex interactions can be deduced by knowing the velocity field the vorticity would induce.

Flow visualization has proven to be a very useful tool for determining the nature of the large-scale structure for any particular flow geometry. It is one of the most straightforward methods of obtaining *field* measurements of the vortex trajectories. Previous stigma associated with doing flow visualization (due to its perceived qualitative nature) and a reliance on time-averaged measurements at individual *points* actually hindered the turbulence research community's realization of the existence of large-scale structures. If the tracer is somehow placed in the cores of the vortices of interest early on in their evolution, then their motion due to convection can be observed.

There has been, since Roshko's (1976) paper, an explosion of work in the area of elucidating large-scale structures in turbulence, particularly the free shear layer problem. This paper will give some examples of how visualization has been used to discover the structure in a limited subset of this research: separated flows, an above-ground nuclear test, water drops impacting on a pool of still water, and bursting air bubbles initially resting on such a pool. Each flow will be discussed briefly in terms of the visualization technique used and an overview of what conclusions about the structure as well as other flow aspects could be drawn from the results. Rather than a comprehensive review of the research done for each flow (for example, for the separation bubble there has been quite a lot as is reviewed in Sigurdson, 1995), we will chronicle the efforts of one research program designed to elucidate the interconnectedness of the large-scale structure between these particular flows.

There is a surprising similarity between the structures that is the focus of the program. These similarities will not be discussed in detail here, other than to explain the contribution of the flow visualization. The similarities will be part of other papers to follow. In an extension of the chemistry analogy suggested earlier, the study of large-scale structures in turbulence is, relatively, only slightly more advanced than the fuzzy understanding of the constituents of matter in Dimitri Mendeleyev's time. Yet Mendeleyev, in 1869, was able to construct a periodic table which predicted possible chemical reactions and the existence of previously unknown elements. Perhaps it will be possible to organize the turbulent flow structures in a "periodic table" in a similar way to Mendeleyev, which may help improve the overall understanding and possibly predict the existence of unknown flow behaviors.

General Methodology

Two main techniques of visualization will be discussed in the following sections: in air, direct smoke injection via probe or smoke-wire, and in liquids, direct injection of dye, either colored or fluorescent. Illumination is primarily from tungsten or strobe lamps for the former, and fluorescent tube banks, strobe or argon-ion laser for the latter. Data acquisition is mainly through a 35-mm camera, however, a 16-mm movie camera and a standard video camera running with continuous illumination for slow flows or synchronized to a strobe for faster flows have also been used to provide sequential information. The video camera synchronized to the strobe is especially useful for immediate feedback on the fine tuning of the experiment. If expense is not a consideration, there are, of course, constantly improving means of electronic image capture, not discussed here. Other than the use of the laser, the techniques used here are not particularly sophisticated, yet we shall see that for the first observation of a flow they can be very informative.

The general methodology in each case is to ensure that some tracer (dye or smoke) is injected at or passes through the region of vorticity creation, or due to the nature of the flow becomes associated with or entrained into the vorticity-bearing parts of the flow. Freymuth (1966) notes that in a two-dimensional inviscid fluid flow, due to the Helmholtz equation, the vorticity is convected in the same way as a material element of the fluid. The concept can be extended to three-dimensional inviscid flow by the application of Helmholtz's laws of vortex motion of 1858. These state that for an ideal barotropic fluid under the action of conservative external body forces, vortex lines move with the fluid (Saffman, 1992). Therefore, if tracer is applied to the fluid material elements that are vortical, the tracer will follow the path of the vorticity. There is not, however, a linear relationship between the *concentration* of dye and the *magnitude* of the vorticity, only a one-to-one relationship between the locations. For instance, vorticity magnitude is increased when stretched, whereas the dye concentration will weaken. This is discussed by Kida and Takaoka (1994).

Complications can also arise due to the fact that fluids are not, in general, inviscid. This means that the Schmidt number (Sc) becomes important, which represents the ratio of the diffusion rate of the vorticity due to viscosity to the rate of diffusion of the tracer. (Sc = ν/k, ν = kinematic viscosity, k = diffusivity of tracer.) As an example, for dye in water the vorticity diffuses faster than the dye, therefore the absence of dye in some location does not guarantee the absence of vorticity there; the vorticity may have diffused to that place while the tracer hasn't. However, if the dye is initially placed with the vorticity, the locations of the dye can still be inferred to be the positions of highest vortical intensity. If the Reynolds number (Re) of the flow is high (indicative of the ratio of inertial forces to viscous forces, Re = $\rho UL/\mu$, ρ = density of fluid, U = characteristic velocity, L = characteristic length, and μ = absolute viscosity), then the convection of the vorticity is more important than the diffusion and the location of tracer becomes more closely associated to the location of vorticity.

Another caution to be heeded is that the tracer does not become canceled in the presence of other tracer which identifies vorticity of opposite sign, whereas it is possible that the vorticity could. Also, if the tracer is *not* initially placed with the vorticity, very careful interpretation of the results must be exercised.

Cimbala et al. (1988) comment on another danger. They studied the streaklines emanating from a smoke-wire which were introduced upstream of the Kármán vortex street behind a circular cylinder at moderately low Reynolds numbers. Previous interpretations indicated vortices, when in actuality the wake had relaminarized. This alerted researchers to the difficulties involved in streakline interpretation and the importance of introducing the tracer near the part of the flow that was of interest. The streaklines represent an integration of all that has happened to the tracer particles from the points of introduction, and *not* only what has happened in the instants before the photograph is taken.

Specific Examples

The Blunt Cylinder and the Blunt Flat Plate

A characteristic three-dimensional, unsteady vortex structure has been proposed for the flow in the reattachment region of a separation bubble (Sigurdson and Roshko, 1984;

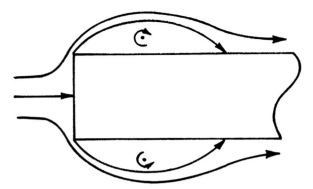

Figure 1 Profile view of the cylinder or plate showing time-averaged streamlines.

Sigurdson, 1986). The bubble was located downstream of the sharp-edged blunt face of a circular cylinder aligned coaxially with the free stream. Another bubble was also studied, the one associated with the two-dimensional analog of this body shape, the sharp-edged blunt flat plate. The same qualitative vortex structure was found to occur there. Figure 1 is a diagram showing the profile view of the cylinder or plate and the time-averaged separating and reattaching streamline that borders the recirculation region.

Flow visualization also played an important role in determining the influence of periodic velocity perturbations on the circular cylinder flow (Sigurdson, 1995; Sigurdson, 1986; Sigurdson et al., 1981). This research was initiated to determine whether the flow could be influenced in a beneficial way, for instance, to reduce the drag on the cylinder. Acoustically generated velocity perturbations were introduced to the flow through a small circumferential gap located immediately downstream of the fixed separation line at the sharp shoulder of the cylinder face. Time-averaged pressure measurements indicated that the drag was reduced. The task then came to explain why and to find what frequencies and amplitudes of excitation created the optimum results.

Smoke was injected into the flow using two methods, from a probe upstream of the model and from a smoke-wire. The smoke-wire is a wire which has had oil dripped down it that beads up into discrete accumulations. When current is passed through the wire it heats up and smoke appears behind the beads to form a series of streaklines. The smoke-wire is described by Corke et al. (1977). Photographs confirmed measurements made with hot-wire anemometers. These indicated that in some ranges of frequency the vortices in the shear layer would "lock in" to the forcing. Photographs also allowed the measurement of the wavelengths of the vortices and it immediately became clear that optimum effects occurred when the vortices were of comparable size to the separation bubble height (Sigurdson et al., 1981). The streaklines in the upstream stagnation region and part of the separation bubble are seen in Figure 2. This form of visualization not only indicates where the turbulent vortices are, but gives an excellent indication of the streamlines in the approximately steady potential flow region distant from the turbulence. The streaklines from the smoke-wire are associated with the streamlines if the flow is steady. Where the flow is *not* steady near the turbulence, this leads to the common novice's misinterpretation of the streaklines there to be streamlines as well. Therefore, far from the turbulence the ratio of vertical velocity v to horizontal velocity u can be calculated from the measured slope of the streaklines. This has been used to comment on the change in entrainment by the turbulent separation bubble with and without acoustic excitation (Sigurdson, 1995; Sigurdson, 1986).

Although there were indications in the smoke-wire results, it was visualization in a low-speed water channel which stimulated several ideas for the flow physics. Various

Figure 2 Streaklines impinging on the flat face of a circular cylinder aligned with the free stream. Reynolds number based on diameter is 21,200.

Lucite models were used and cobalt blue and fluorescent dye were injected in many different ways. Laser-induced fluorescence (LIF) was used to visualize a planar section of the circular cylinder flow. The argon-ion laser beam was optically processed to form a thin sheet which passed through the translucent model along a diameter. Fluorescent dye was injected from a probe upstream or from two ports located in the plane of the laser sheet diametrically opposite each other on the cylinder face. By comparing the results it was confirmed that the injection from the ports made no qualitative difference to the flow behavior, providing the injection flow rate was low enough. In all cases of injection mentioned here, the dye could be turned off and there would be enough dye resident in the recirculating region of the bubble to see that there was no noticeable effect.

Figure 3 shows one of the results for an unforced flow. The primary advantage of the LIF technique is to be able to exclude visualization of the dye which is outside the plane of the laser sheet. In this case, the vorticity of interest is primarily created on the flat face of the cylinder. If dye is placed there it will track the vorticity as discussed in the previous section. The conclusion from this photograph and others was that there are large vortical

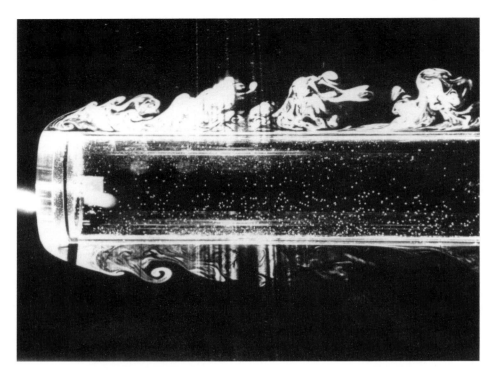

Figure 3 Laser-induced fluorescence visualization of the flow over a circular cylinder. Reynolds number based on diameter is 5150. (From Sigurdson, L.W., *JFM*, 298, 139, 1995; Sigurdson, L.W., Ph.D. thesis, Aeronautics Department, California Institute of Technology, 1986.)

structures emerging from the separation bubble and convecting downstream (Sigurdson, 1995; Sigurdson, 1986; Sigurdson and Roshko, 1985). They bore a resemblance to the well-known periodic Kármán vortices shed behind a circular cylinder aligned *transverse* to the free stream. This led to a hypothesis that there was a similarity between this "shedding" instability and Kármán vortex shedding. Now, however, the real vortices were interacting with their images due to the presence of the wall, rather than the real vortices interacting with other real vortices as in the Kármán vortex shedding case.

Further consideration resulted in a method of scaling the frequency of excitation in a manner similar to that known to be true for Kármán vortex shedding. The optimum forcing frequencies were ones which were amplified by this "shedding" type of instability. Measurements of vortex wavelength from photographs were useful in confirming these hypotheses. Estimates of the frequency could be made by using the measured wavelengths and estimated convection velocity. Video and film sequences were used for this as well. Flow visualization also played a part in validating a theory to estimate the trajectory of a free streamline which separates from the blunt face of the cylinder.

One of the things which slowed the realization of the Kármán shedding analogy was that the structures were not always so clearly apparent as in Figure 3. Further experimentation with back-lit cobalt blue dye helped explain why. The key was to introduce the dye along a slit immediately behind the sharp edge of the face of the translucent cylinder. The slit was arranged to extend along one half of the entire circumference of the cylinder. This would more comprehensively mark the vorticity and gave a three-dimensional view of the turbulence, rather than the strictly two-dimensional information from the LIF results. An example of the results is provided in Figure 4. The initial free shear layer structures are primarily two-dimensional but evolve into boundary layer-type structures as they near attachment and interact with the wall (Sigurdson and Roshko, 1984; Sigurdson, 1986).

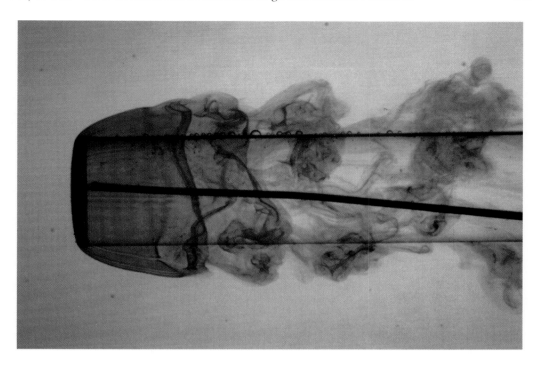

Figure 4 Cobalt blue dye injected from a slit near the leading edge of a cylinder aligned with the free stream. Reynolds number based on diameter is 2750. (From Sigurdson, L.W., Ph.D. thesis, Aeronautics Department, California Institute of Technology, 1986.)

Some segments form what roughly appear to be hairpin "loops" which convect away from the wall and downstream. Spanwise adjacent segments convect toward the wall and upstream. The loops are sometimes clearly arranged in a staggered pattern (also more recently noted by Sasaki and Kiya, 1991). A line vortex model was presented by Sigurdson (Sigurdson and Roshko, 1984; Sigurdson, 1986) that explains these observations. It was the basic three-dimensionality of the characteristic large-scale structure that complicated interpretation of the planar LIF photographs, where often the structures could not be clearly seen.

The two-dimensional counterpart of the blunt-faced cylinder, the blunt flat plate was investigated next to allow better visualization of the loop structures. Again dye was injected from a slit several plate thicknesses in length immediately behind the separation line, and at times also from a similar slit located near the reattachment region. It confirmed that the structures were qualitatively similar to those for the cylinder. Fluorescent dye was illuminated with a laser in two ways: perpendicular to both the wall and the free stream (Figure 5) and parallel to the wall (Figure 6). Figure 5 shows the characteristic "mushroom" shapes associated with the two legs of the vortex loops (the lower images are a reflecton in the plate). Figure 6 shows a staggered pattern formed by the structures as they rise away from the plate and intersect the laser sheet. Conclusions concerning the evolving topological nature of the vortex structures and their wavelengths could be drawn from photographs like these. The proposed structure reconciled apparently contradictory prepositions concerning the fate of the structures as they encounter reattachment (Sigurdson and Roshko, 1984; Sigurdson, 1986).

The nature of this flow is one of a recirculation region (separation bubble) which has, *on average*, a set of closed streamlines bounding it (Figure 1). At any *instant*, however, the three-dimensional hairpin structures were escaping the trapped orbits of the recirculation region and convecting downstream. The legs of the hairpin loop would reach upstream

Figure 5 LIF view looking upstream, sheet perpendicular to both plate and free stream, located 8.3 plate thicknesses downstream of separation. Reynolds number based on plate thickness is 890.

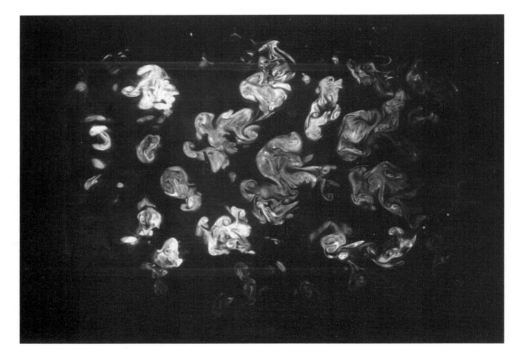

Figure 6 LIF plan view, sheet parallel to the plate, located 0.9 plate thicknesses away from the plate wall. Flow from left to right, left edge of photograph is plate leading edge. Reynolds number based on plate thickness is 850. (From Sigurdson, L.W., Ph.D. thesis, Aeronautics Department, California Institute of Technology, 1986.)

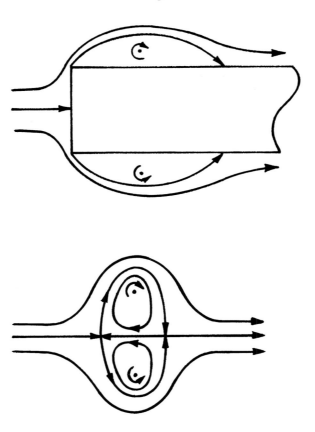

Figure 7 Comparison of the time-averaged streamlines for the cylinder or plate (top) to those for a vortex ring (bottom).

into the recirculation region and remain joined there to the adjacent loops' legs. At a later time these upstream loops would be expected to escape as well. This is explained in detail in Sigurdson (1986). The important image to hold on to now is that of the loops escaping the trapped streamlines, for it is similar to the next flows we will discuss.

Impacting Water Drops and Above-Ground Nuclear Tests

Knowledge of the three-dimensional loop structure described in the last section allowed recognition of its topologically similar existence in several other flows. Two of these are the largest-scale structure of a water drop impacting a pool of still water and the mushroom cloud from an above-ground nuclear blast. Both flows contain primary vortex rings which have similar time-averaged streamlines to the cylinder or plate flow as shown in Figure 7. The geometry of the streamlines for a vortex ring can be obtained by shrinking the cylinder to infinitesimal diameter.

First the characteristic shape of the vortex loops was seen in a bomb photograph taken by the U.S. Department of Energy in 1957, shown in Figure 8 on the right. Then aided with the knowledge that vortex tubes cannot end in the fluid (Saffman, 1992), and what kind of velocity field is implied by the presence of the vorticity through the Biot-Savart law, a hypothesized vortex skeleton structure was devised to explain the observed shapes (Sigurdson, 1987). The tracer added to the flow in this case is primarily heat and dust. The heat is from the original fireball which for this particular test (the "Priscilla" event of Operation Plumbob) was large enough to reach the ground from where the bomb was initially suspended.

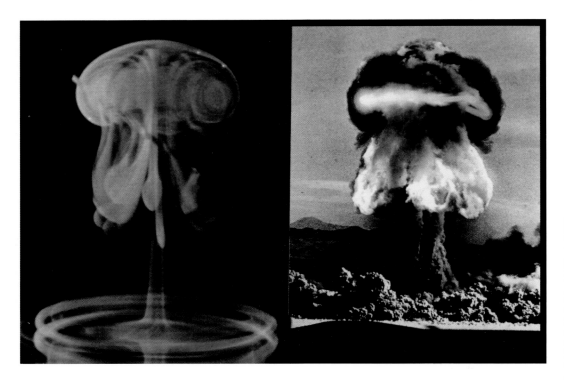

Figure 8 Comparison of the vortex structure created by a 2.6-mm water drop 50 ms after impacting a pool of water after a fall of 35 mm (inverted, left) and an above-ground nuclear test, Nevada, 1957, U.S. Department of Energy (right). (Left, from Peck, B. and Sigurdson, L.W., *Phys. Fluids*, 6[2] [Part 1], 564, 1994.)

It is conjectured that the vorticity in this problem is generated initially primarily at the interface between the hot fireball gas and the cooler surrounding air. This causes a density gradient which creates vorticity when acted on by the pressure gradients. The pressure gradients are associated with the shock wave reflected upward from the ground and the vertical hydrostatic pressure gradient from the usual resting atmosphere. This type of vorticity creation is sometimes called baroclinic torque. Since the heat tracer is not *only* initially located at the interface but throughout the fireball, the presence of heat does not in itself imply the presence of vorticity as in the previous examples. The ramifications of this to photograph interpretation concerning the water drop experiments will be discussed more fully later.

The line vortex model consists of closed line vortices in a similar fashion to the flow in the previous section: the primary ring with four azimuthal waves, four connected loops reaching to the primary ring, and four counterrotating vortex pairs forming a "stalk" which reaches from the primary ring to another ring of opposite sign situated at ground level. The topological similarity to the separation bubble is in the way the hairpin loops reach into the trapped orbits of the primary ring which are analogous to those of the separation bubble.

The detailed similarity between the water drop and the bomb was first noted by the author while looking at the drop photographs of Okabe and Inoue (1961). It seemed that the drop had a very similar basic line vortex structure to the bomb. Some results of this comparison and reasons for this similarity have been discussed previously (Sigurdson, 1987; Sigurdson, 1991), and will be the subject of another paper. Although many conclusions could be drawn from these photographs alone, more detailed information about the drop was necessary to confirm the hypothesized structure. An experimental program was begun with Bill Peck to shed more detailed light on the structure of the impacting drop. Details are given in Peck and Sigurdson (1994) which confirm the basic hypothesized structure above and add greatly to the deeper understanding of its birth and evolution. The number of hairpin loops for the drop was observed to be three, four, or five. A photograph of the impacting drop-created vortex structure taken in our laboratory (for very specific conditions) is compared to the blast-generated vortex structure in Figure 8.

The 2.6-mm drop was dyed with fluorescent tracer and illuminated by a strobe flash 50 ms after it had fallen 35 mm and impacted the clear pool. Interruption of a laser beam by the falling drop was used to trigger a timer which controlled the time of the strobe flash after impact of the drop. Details of the careful experimental procedure necessary to create these drop photographs is given by Peck and Sigurdson (1994) and Peck et al. (1995). Various orthogonal camera angles were used to observe the structure.

Interpretation of where the vorticity is becomes more complex in this experiment because it was the *entire* drop which was originally carrying the dye, and the entire drop fluid at early times after impact is not expected to be vortical. It is expected that most of the vorticity is created at the air–water interface that is associated with the impact, therefore ideally the dye would be placed only there to track that vorticity. Consequently, on interpreting these photographs conclusions are drawn based largely on how the dye belonging to the bulk of the drop is being convected around by the vorticity. As much weight as in the cylinder or plate experiment cannot be given to the simple presence of dye as an indication of presence of vorticity. Figure 9 from Peck and Sigurdson (1991) shows a drop impacting with approximately the same conditions as in Figure 8, but at an earlier time, 9 ms after impact rather than 50 ms. This symmetric structure can evolve into the

Figure 9 Structure created by a 2.8-mm water drop 9 ms after impacting a pool of clear water. It had fallen 38 mm. (From Peck, B. and Sigurdson, L.W., *Phys. Fluids*, 34[9], 2032, 1991.)

structure shown in Figure 8. The presence of vorticity is indicated by the rolling up of the dye at the edges. The photograph is taken from slightly below the water surface so a reflection of the dye is evident in the upper part. At this time the impact crater that is temporarily formed in the surface is present.

Through the use of visualization for this flow several conclusions could be drawn. Besides the structural information discussed earlier, information was learned about the transition from laminar to turbulent flow and back again (Sigurdson and Peck, 1995). Photographs recorded the shape of the oscillating drop as it falls and the capillary and larger surface waves that occur during the impact. It was found that air becomes trapped beneath the drop to sometimes form a single bubble. Velocity measurements of the resulting vortex ring could be made by making multiple exposure photographs with known time delays between strobe flashes (Peck and Sigurdson, 1995).

Bursting Air Bubbles

If a falling water drop impacting the free surface of a pool of water creates a vortex ring in the water beneath it, then one might presume the inverse of this problem would result analogously. This would be an air bubble rising in a pool of water and penetrating the free surface to create a vortex ring in the air above it, the geometry shown in Figure 10. In order to verify this and provide information about the birth and evolution of the vortex structures resulting, experiments were designed with James Buchholz to study a simpler version of the problem. It consisted of a smoke-filled air bubble resting on a free surface which was subsequently forced to break by the transmission of an electrical spark across the bubble wall. Control of the spark timing allowed photographs to be taken at various times after bubble breakage. Figure 11 from Buchholz et al. (1995a) is a triple exposure. One strobe flash exposure shows the fluorescent dyed bubble liquid before breakage. One exposure is from the flash of the spark itself, and one exposure by the strobe again shows the resulting primary vortex ring. One spark electrode can be seen coming down from the top of the bubble and another from below the bubble film. Details of the apparatus can be found in Buchholz et al. (1995b). Although researchers as far back as Rogers in 1858 (Rogers, 1858) had observed similar vortex rings, this experiment is apparently the first one to produce a photograph of the vortex ring structure.

Interpretation of the photographs is similar to the drop examples; care must be employed because the smoke tracer is originally dispersed approximately uniformly within the air bubble and not simply at the liquid–air interface where the vorticity is created. Preliminary photographs (Sigurdson et al., 1995) show that the initial roll-up of the vorticity into the primary vortex involves many smaller discrete vortex rings, a result of the Kelvin-Helmholtz instability of a free shear layer. There appear to be loops left in the wake of the primary vortex with some similarity to the drop structure, but details are still forthcoming. One result concerning the vorticity creation is that there appears to be

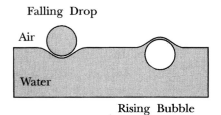

Figure 10 Schematic comparison of the falling drop and the rising bubble.

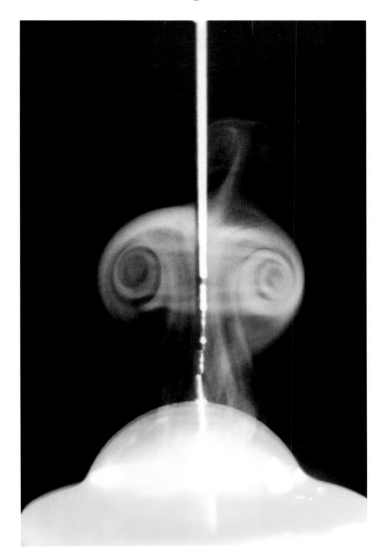

Figure 11 Triple exposure of a smoke-filled bursting air bubble showing the original bubble and resulting vortex ring. Base width of the bubble is 1.54 cm. (From Buchholz, J., Sigurdson, L.W., and Peck, B., *Phys. Fluids*, 7[9], S3, 1995.)

appreciable vorticity of opposite sign to the primary ring that is entrained into it (Sigurdson and Buchholz, 1995). Any model for vorticity creation at an air–water interface must be able to predict this and account for the fact that vortex rings are generated for both the bubble and drop despite the fact that the density and viscosity gradients across the interface are reversed in the two cases. As with the drop, velocities of the primary vortex ring can be measured using multiple-strobed photographs.

Conclusions

Four different flow types have been discussed in terms of how flow visualization has been used to elucidate the large-scale structure in each. A general methodology has been given to use for determination of the structure from interpretation of photographs obtained by direct injection of tracer materials. Some danger areas of misinterpretation were discussed. For each flow type the basic nature of the structure was described.

Concerning the flow visualization techniques themselves, it is interesting to note how successful straightforward application of direct injection of tracers can be in providing some initial understanding of the flow behaviors. Once some idea of the nature of the structure is obtained in this way, more quantitative measurements can be much more wisely designed to characterize the flows. Although in the separated flows discussed here there was already a significant body of measurements to guide researchers, in the other examples there was not, and the contribution of the flow visualization data is quite valuable. In the separated-flows case, flow visualization offered new ideas for interpretation of previous results and resulted in useful physical insights. Some quantitative measurements of other flow features were discussed, for example, vortex ring convection velocity. When the details of the velocity field are required, elaborate methods are now available.

Concerning the nature of the flow structures observed in these examples, only a brief outline has been given. A much more in-depth consideration of the apparent similarity between the three-dimensional topology of the vortex structures discussed will be given in another paper. In these flows the topological arrangement of hairpin loops extending into the time-averaged trapped orbits of a recirculation region is a common theme. Knowledge of one led to the next. Many other flows exhibit these similarities (Sigurdson, 1986). Yet, at present, the only way to determine the structure is experimentally or through difficult (often still impossible) computation, and then careful interpretation of results using the conservation laws and knowledge of vortex dynamics. It is hoped that presenting the flow information all together in this paper may inspire other researchers to seek the equivalent of Mendeleyev's "periodic table" for turbulent structures which would aid in this process.

Acknowledgments

The work on the separated flow structure was done while I was a graduate student with Dr. Anatol Roshko and he is thanked for sharing his wisdom and support. That work was supported by the Office of Naval Research Contract N00014-76-C-0260. Some of the work on the comparison of the water drop and atomic test was supported by Darpa URI N00014-86-K-0758. The remaining work has been funded by the Natural Sciences and Engineering Research Council of Canada grant OGP0041747. Bill Peck, James Buchholz, and Bernie Faulkner are thanked for their critical roles in obtaining the drop and bubble photographs included here, among other things.

References

Buchholz, J., Sigurdson, L. W., and Peck, B., Bursting soap bubble, *Phys. Fluids*, 7(9), S3, 1995a.
Buchholz, J. H. J., Sigurdson, L. W., and Peck, B. J., An apparatus to study vortex rings emerging from bursting bubbles, in *Proc. 7th Int. Symp. Flow Vis.*, Seattle, WA, September 11–14, 1995b, 146.
Cimbala, J. M., Nagib, H. M., and Roshko, A., Large structure in the far wakes of two-dimensional bluff bodies, *JFM*, 190, 265, 1988.
Corke, T., Koga, D., Drubka, R., and Nagib, H., A new technique for introducing controlled sheets of streaklines in wind tunnels, *IEEE Publ.*, 77-CH 1251-8 AES, 1977.
Freymuth, P., On transition in a separated laminar boundary layer, *JFM*, 25(4), 683, 1966.
Kida, S. and Takaoka, M., Vortex reconnection, *Annu. Rev. Fluid Mech.*, 169, 1994.
Okabe, J. and Inoue, S., The generation of vortex rings. II, *Rep. Res. Inst. Appl. Mech. Kyushu Univ.*, 9(36), 147, 1961.
Peck, B. and Sigurdson, L. W., Impacting water drop, *Phys. Fluids*, 34(9), 2032, 1991.

Peck, B. and Sigurdson, L.W., The three-dimensional vortex structure of an impacting water drop, *Phys. Fluids*, 6(2) (Part 1), 564, 1994.

Peck, B. and Sigurdson, L. W., Vortex ring velocity resulting from an impacting water drop, *Exp. Fluids*, 18, 351, 1995.

Peck, B., Sigurdson, L. W., Faulkner, B., and Buttar, I., An apparatus to study drop-formed vortex rings, *Meas. Sci. Tech.*, 6, 1538, 1995.

Rogers, W. B., On the formulation of rotating rings by air and liquids under certain conditions of discharge, *Am. J. Sci.*, 26, 246, 1858.

Roshko, A., Structure of turbulent shear flows: a new look, *AIAA J.*, 14(10), 1349, 1976.

Saffman, P. G., Vortex interactions and coherent structures in turbulence, in *Proc. Symp. on Transition and Turbulence*, Academic Press, New York, 1980, 149.

Saffman, P. G., *Vortex Dynamics*, Cambridge Monographs on Mechanics and Applied Mathematics, Batchelor, G. K. and Freund, L. B., General Eds., Cambridge University Press, New York, 1992.

Sasaki, K. and Kiya, M., Three-dimensional vortex structure in a leading-edge separation bubble at moderate Reynolds numbers, *Trans. ASME I: J. Fluids Eng.*, 113, 405, 1991.

Sigurdson, L. W., (1986) The structure and control of a turbulent reattaching flow, Ph.D thesis, Aeronautics Department of California Institute of Technology, 1986.

Sigurdson, L. W., Three-dimensional vortex structure of the starting vortex ring, *Bull. Am. Phys. Soc.*, 32, 10, 1987.

Sigurdson, L. W., Atom-bomb/water drop, *Phys. Fluids A*, 3(9), 2034, 1991.

Sigurdson, L. W., The structure and control of a turbulent reattaching flow, *JFM*, 298, 139, 1995.

Sigurdson, L. W. and Buchholz, J. H., The vortex structure of impacting water drops and bursting air bubbles, *Bull. Am. Phys. Soc.*, 40(9), 1920, 1995.

Sigurdson, L. W. and Peck, B., Three-dimensional transition of the vorticity created by an impacting water drop, in *Advances in Turbulence V, Proc. 5th European Turb. Conf.*, Siena, Italy, July 5–8, 1994, Benzi, R., Ed., Kluwer Academic Publications, Dordrecht, 1995, 470.

Sigurdson, L. W. and Roshko, A., The large-scale structure of a turbulent reattaching flow, *Bull. Am. Phys. Soc.*, 29(9), 1542, 1984.

Sigurdson, L. W. and Roshko, A., A controlled unsteady excitation of a reattaching flow, *AIAA Paper*, 85-0522, 1985.

Sigurdson, L. W., Cimbala, J., and Roshko, A., Controlled excitation of a separated flow, *Bull. Am. Phys. Soc.*, 26, 9, 1981.

Sigurdson, L. W., Buchholz, J. H., and Peck, B. J., The vortex rings created by bursting air bubbles and impacting water drops, in *Proc. 15th Can. Cong. Appl. Mech. (CANCAM)*, Vol. II, University of Victoria, Victoria, BC, CA, May 28–June 2, 1995, 542.

chapter seven

Analysis of Turbulence and Vortex Structures by Flow Mapping

C.A. Greated, C.E. Damm, and J. Whale

Department of Physics and Astronomy, The University of Edinburgh

Abstract—*The technique of particle image velocimetry (PIV) flow mapping is reviewed and comparisons made with laser Doppler anemometry (LDA). Results are presented showing the application of PIV to the determination of coherent structures in grid-generated turbulence and theoretical expressions are presented for the errors associated with the computation of statistical parameters. Measurements are also presented showing the vortex structure in the wake of a model wind turbine. These studies have revealed fundamental inadequacies in existing computer codes used by the wind turbine industry.*

Nomenclature and Abbreviations

$a(x,y)$	Optical intensity
$A(x,y)$	Autocorrelation function
BEMT	Blade-element/momentum theory
c	Chord length of blade
CCD	Charge coupled device
CW	Continuous wave
d_i	Image diameter on film or CCD array
d_p	Seeding particle diameter
d_s	Diffraction-limited spot diameter
d_r	Point spread function on film or CCD array
F	Fourier transform
F^{-1}	Inverse Fourier transform
FFT	Fast Fourier transform
K	Normalized autocovariance function
ℓ	Grid bar diameter
L	Grid bar spacing
LDA	Laser Doppler anemometry
M	Magnification
MSE	Mean square error
N1, N2	Size of sample area measured in numbers of pixels
R	Rotor radius

Re	Reynolds number
rms	Root mean square
SLM	Spatial light modulator
t, t_o, τ	Time coordinates
T	Averaging time
U, Uo	Flow velocity
V	Flow velocity in direction perpendicular to U component
W	Local velocity at the blade
X,Y	Spatial coordinates
α	Parameter defining width of spatial correlation function
η, ξ	Spatial coordinates
λ	Tip speed ratio, i.e., ratio of blade tip-to-flow velocity
σ	rms velocity fluctuation
ν	Kinematic viscosity
ω	Angular speed of rotation

Evolution of the PIV Flow Mapping Technique

Nonintrusive flow measuring techniques are becoming an increasingly familiar part of the fluid dynamics laboratory and industrial test facility. Since its inception in 1966, laser Doppler anemometry (LDA) has undergone continuous development (Durrani and Greated, 1977). Aided by developments in other technologies, notably in lasers, fiber optics, and computing, it is now a most useful tool for measuring under a wide range of conditions, for example, in high speed wind tunnels, microscopic-scale biological flows, combustion rigs, and two-phase flows. Despite its great success, however, LDA is fundamentally a point measuring technique. The time evolution of flow velocity can be measured with great accuracy at a point, but if a map of the area is to be obtained, it must be built up point by point; the implication here is that the flow must be steady and accurately repeatable. Particle image velocimetry (PIV), on the other hand, provides a quantitative map of instantaneous flow velocity over a large field (Gray et al., 1991).

PIV is sometimes compared to streak photography, where the path lengths of individual marker particles in the flow are measured for a given exposure time. Streak photography, however, suffers two main disadvantages. First, only a few marker particles can be used, otherwise the paths overlap too often and the individual paths cannot be separated. Second, when used with sheet illumination, the particles move out of the measurement plane, artificially shortening the streak lengths and causing errors in the velocity measurement.

Light Sheet Formation

With PIV, a two-dimensional sheet in the flow, seeded with small marker particles, is illuminated stroboscopically. A double, or multiple, exposure photograph of this plane is then taken. The spacing between the images of each particle on the film gives the local velocity. This photograph is then analyzed to obtain the complete set of velocity vectors over a grid of points covering the whole field. Two different techniques are commonly used for producing the stroboscopic light sheet. In the first of these a continuous wave (CW) laser is deflected off a multifacet mirror so that the beam is scanned through the measurement area. The shutter time on the camera is normally set to five or six times the period of the beam scan, in order that this number of exposures is recorded on each negative. In the second technique a pulsed laser, typically a NdYag, is used to produce double pulses, a cylindrical lens being used to spread the beam into a sheet.

A comparison of the techniques (Gray et al., 1991) shows that the scanning beam approach is usually better for flows up to about 10 ms^{-1}. This method gives even illumination over large areas and has the advantage that more than two images of each particle can easily be obtained, thus improving the signal-to-noise ratio in the subsequent analysis. For measurement regions of the order of 1 m^2 and flow velocities of up to 2 ms^{-1}, a 10-W argon ion laser is suitable. The upper limit of velocities that can be measured with the scanning beam system can be increased by, for example, reducing the width of the scanned area. Above about 10 m/s it is necessary to use a pulsed laser. In this case the energy per pulse is almost constant, so that measuring higher velocities does not imply a proportionate increase in laser power, as it does with the scanning beam. With the pulsed laser it is more difficult to obtain a uniform illumination over a large area. The number of pulses per exposure can be increased from two to four by coupling two lasers, but this is a costly option.

Image Recording

Conventional photography has been the most commonly used method of obtaining PIV images. It gives very high resolution (typically, film can resolve about 200 lines per millimeter) and a ready means of data storage. A high quality flat field lens is required to minimize distortion but errors are introduced due to the parallax effect if the out-of-plane velocity component is high. These errors may be reduced by choosing a long focal length imaging lens, but this advantage must be offset by the greater difficulty of obtaining a sharp focus and increased susceptibility to vibration.

The illumination interval, i.e., the strobe period, is set such that the estimated maximum velocity in the flow gives a particle separation on the film which is just resolvable by the analysis system. Once this has been chosen, the shutter speed can be set to give the required number of exposures. Choice of the optimum lens aperture is also important. It should be remembered that the largest aperture (small f number) gives the poorest depth of field, making focusing difficult. Achieving a sharp focus is vital in the PIV process. On the other hand, small apertures (large f numbers) may result in the particle images being diffraction limited and hence artificially enlarged in size. A good compromise is usually f/4 or f/5.6.

Image Shifting

One of the most serious limitations of PIV in its simplest form is the fact that it does not resolve the direction of the flow, i.e., from the sequence of exposures from any individual particle it is not possible to distinguish which one occurred first and which one last in the sequence. Various techniques have been devised for marking the exposures, for example, by using different colors or by making the first or last pulse in a sequence of different length than the others. However, these have not generally been very successful since the subsequent analysis procedure becomes too complicated. An equally important limitation is the fact that the technique cannot resolve very small velocities (relative to the maximum velocity in the flow), since in this case successive images of individual seeding particles overlap, making accurate determination of the particle separation impossible.

The same directional ambiguity and small velocity problems arise in LDA where they are normally overcome by frequency shifting one of the probing laser beams so that all the velocities are raised above a pedestal value, chosen to be slightly larger than the largest estimated negative velocity in the flow. This pedestal value is subtracted in the final stage of signal analysis. A similar procedure may be used with PIV. An image shift is introduced by panning the camera over the flow at a constant rate, or alternatively by rotating a mirror in front of the camera lens. In this way the image is moved at a constant rate across the image plane and analysis of a static flow produces a constant velocity flow field. It is possible to apply the image shift in different directions, the choice being dependent on the particular flow field under study. If the velocity components in orthogonal directions are of greatly different magnitude, then image shifting in the direction of the smaller component is to be preferred, since the magnitude of the shift required in order to avoid directional ambiguity will be smaller. An example of this might be the oscillatory boundary layer over a flat plate where the shift would be applied in the direction perpendicular to the plate.

Image shifting by panning camera or rotating mirror introduces an error in the velocity measurement, due to the fact that the photographs are taken when the object plane is not precisely perpendicular to the optical axis and the angles between these are different for the successive exposures. This means that the same amount of movement of a particle in the center and edges of the picture shows up as different movements on the image plane. Hence the superimposed velocity is not precisely constant across the whole extent of the image. This effect can be minimized by using a relatively long focal length lens or, alternatively, it is quite simple to use the photographs of static flow to apply a correction in the analysis procedure.

Electro-optic systems for image shifting have been devised (Landreth and Adrian, 1988). These normally make use of a calcite crystal which has the property that it will transmit rays polarized in different directions along different paths. By using this in conjunction with a Pockels cell it is possible to arrange for an image shift without introducing any mechanical components.

Image Analysis

Analysis of the photographic images is one of the most complicated aspects of the PIV technique. The procedures used fall into two categories, particle tracking and correlation (Gray, 1989).

Particle tracking implies that individual particles can be traced in the flow and it is therefore only generally applicable to flows where the seeding is very sparse. This, in turn, implies that only a very sparse array of velocity vectors will be obtained. Nevertheless, the technique may be extremely powerful in certain situations, particularly with high speed air flows or where the characteristics of particle dispersion in a flow are required. PIV photographs can, of course, be scanned manually, but this can be an extremely time-consuming procedure. Various computer algorithms have therefore been devised in order to match up corresponding particle images in the flow.

The usual procedure for analyzing images is autocorrelation (Meinhart et al., 1992). The image (usually a photographic negative) is divided up into a set of small interrogation areas and a local velocity vector is determined for each region. The two-dimensional autocorrelation function of the intensity across the probe area produces two peaks which give the mean displacement of the particles within the area. As with LDA, an additional peak occurs centered at zero, corresponding to the self-correlation of particles; this is normally just discarded in the subsequent analysis. If more than two exposures are recorded then additional higher harmonic peaks are also produced, giving, for example, the correlation between the first and third particles in a sequence.

The autocorrelation function for any particular probe area can most efficiently be obtained by applying two consecutive Fourier transforms. Thus, if $a(x,y)$ is the intensity distribution across a chosen position on the film, F is the Fourier transform and A the autocorrelation function:

$$A(x,y) = F^{-1}\{|F[a(x,y)]|^2\} \qquad (1)$$

In practice the correlation peaks are slightly broadened due to the fact that the probe area is weighted by the intensity profile across the probing beam, but this does not generally affect the velocity measurement.

Fast Fourier transform routines are computationally very efficient, but, even with the fastest computer, the computation time can be very considerable. This problem can be overcome by using optical Fourier transform techniques but at the expense of some loss of flexibility. Three types of analyses are possible.

1. Both Fourier transforms can be computed digitally. In this case it is usual to illuminate the interrogation areas with white light. This type of analysis is the most time-consuming, but, by working in the image plane, boundaries can easily be identified. The type of weighting used to define the boundaries of the interrogation area can also be varied at will.
2. The first Fourier transform may be obtained optically and the second one digitally (Gray and Greated, 1988). Here, a low power laser is used for the probing beam and a lens is used to form the first transform, the probing beam being scanned automatically in small steps across the negative. The squared modulus of the Fourier transform, i.e., the intensity distribution, is in the form of a set of Young's fringes which are captured on a CCD camera positioned at a distance of one focal length behind the lens. This arrangement is quite straightforward to implement and offers a useful increase in processing speed, compared to a fully digital analysis. One problem that needs to be addressed is the elimination of the halo function, associated with the intensity distribution across individual particles, analogous to the low frequency component in LDA. A routine is usually employed for subtracting this, in order that the peaks themselves are not swamped. However, image shifting solves the problem of separating the correlation peaks, since it moves them further from the center of the correlogram.
3. Both Fourier transforms may be generated optically (Jakobsen et al., 1995). The PIV negative is illuminated with a low power laser beam as in the hybrid technique (no. 2 above), but the CCD camera is replaced by an optically addressed spatial light modulator (SLM). While one side of the SLM is illuminated with the fringe pattern, the other side is illuminated with a second coherent beam. In this way the second beam is modulated by the fringe function. A second lens is then used to transform the fringe pattern to the correlation function. The correlation function itself is captured by a CCD array and a search is used to locate the positions of the displacement peaks and hence the velocity vector. With this approach the speed of the Fourier transforms is so fast that the analysis time is limited only by the peak search routine and the time required to move the stepping motor on the transverse on to the next point.

In practice it is necessary to incorporate some form of validation procedure into the analysis to ensure that each velocity vector calculated from a correlogram is a true representation of the real flow. Validation is most readily achieved by comparing the height of the correlation peaks used for measuring the velocity with the rms height of the surrounding correlogram. This gives an effective signal-to-noise ratio which must always be above some prescribed value if the reading is to be taken as valid. It should be noted here that the required correlation peaks will always be set in a noisy background due to the correlations which occur between all of the different particles across the field. Further noise then arises from imperfections in the system, for example, graininess of the film and random variations in the illumination intensity.

CCD Systems

With a CCD-based (charge coupled device) system (Willert and Gharib, 1993) the complete analysis process is automated and the time between recording a PIV image and receiving the flow map is greatly reduced. Using two CCD cameras and a beam splitter, it is possible to capture successive images separated by a very small time interval. Cross-correlation of the two images then yields the velocity field without directional ambiguity, the computations being carried out using digital FFT routines. Using cross-correlation as opposed to autocorrelation not only avoids problems of directional ambiguity, but it improves the effective spatial resolution and dynamic range of the instrument.

Modestly priced CCD cameras are now available with a resolution of the order 1000 × 1000 pixels, comparable to a 35-mm film camera. Higher resolution CCDs can also be obtained but their frame rate is generally rather low, typically of the order of one frame per second. The merits of the CCD system are such that nearly all new PIV systems now employ this approach.

Choice of System Parameters

The choice of optical settings for PIV experiments can be a difficult process due to the interdependence of the parameters. Table 1 shows typical values of parameters used in the Edinburgh experiments (Skyner, 1992), together with the way in which they affect the exposure of the film or CCD array.

Table 1   Exposure Dependence of PIV Parameters

Parameter	Order of parameters	Exposure dependence (exponent of proportionality)
Laser power	10 W	1
Illumination interval	1 ms	1
Camera distance	1 m	−2
Focal length of lens	1 m	0
Aperture (f-number)	f4	−2
Film speed (ASA)	400	1
Laser sheet length	1 m	−1
Laser sheet thickness	1 mm	−1
Particle diameter	10 μm	3

It is important to ensure that the average separation of particle images is significantly greater than the particle image diameter, otherwise successive images, and hence the peaks on the correlogram, will overlap. The image diameter on the film or CCD array is given by (Adrian, 1988)

$$d_i = \left(M d_p^2 + d_s^2 + d_r^2\right)^{1/2} \qquad (2)$$

where
- d_p = the seeding particle diameter
- d_s = the diffraction limited spot diameter of the optical system
- d_r = the point spread function of the film emulsion or CCD array, which can be approximately by the width of a line which is just resolved
- M = the magnification

The errors that occur in the recording of individual velocity vectors have been well researched and are summarized in Gray (1989). In essence, these can be classified as either random or systematic. Systematic errors occur, for example, from the seeding particle dynamics, i.e., the fact that the particles do not precisely follow the fluid motions, and the existence of velocity gradients within the individual interrogation regions. Systematic errors may also occur due to optical distortions and misalignments. In the tests described here, these were corrected by first conducting tests in still fluid and then subtracting these results from the flow records. Random errors are principally due to the unknown phase of the seeding particles within the interrogation regions which give rise to a noise background on the correlogram and an uncertainty in the detection of the correlation peaks. In highly three-dimensional flows, such as the ones for which results are presented in this paper, the out-of-plane components may also introduce a random error caused by parallax effects in the photographic imaging. These are only significant, however, near the edges of the flow field. The random errors are estimated to be of the order 1% for the parameters used in the experiments presented here.

Measurement of Turbulence Structure

One of the major applications of PIV flow mapping is in the measurement of turbulent flows. Traditionally, point measuring techniques have concentrated on the extraction of turbulence statistics from time records of velocity, e.g., mean velocity, rms turbulence level, and velocity correlations. With PIV the problem is to extract turbulence parameters from a discrete series of spatial velocity maps.

Turbulence Experiments

The results presented here are for experiments carried out in a small wind tunnel. Air from a blower fan passed through a series of mesh screens, followed by a settling chamber, and then into a 12:1 area ratio contraction section, before entering the 52-mm^2 working section. A rectangular grid of bars was located at the upstream end of the working section, in order to produce an approximately homogeneous turbulence field. Grid-generated turbulence has been used extensively as a good approximation to homogeneous isotropic turbulence (Batchelor, 1986). The grid bar diameter was $\ell = 2.44$ mm and the grid spacing was L = 8.41 mm, corresponding to a solidity factor of $\ell(2 - 1/L)/L = 0.496$. The mean flow velocity was 2 m/s, giving a grid Reynolds number of 1120.

For these tests, scanning beam illumination was used, with an 18-facet rotating mirror and the mean flow direction in the plane of the light sheet. A 15-W Argon-ion continuous wave laser acted as the light source, radiating with wavelength 400 to 530 nm, i.e., green/blue. The walls of the test section were all made of glass, in order to allow complete optical access. Photographs of the flow were recorded with a Nikon F801 film camera fitted with an AF Micro-Nikkor 60-mm lens, the optical axis being perpendicular to the light sheet. The Tmax 400 film used had a resolution of 100 lines/mm. An aerosol of corn oil, with a mean droplet diameter of 1 µm, was used for seeding. The mean particle image size was calculated from Equation 2 to be d = 15 µm, which was acceptably small since the mean particle image separation was 0.16 mm, i.e., an order of magnitude greater.

A typical velocity map obtained by PIV is shown in Figure 1. This is an instantaneous picture of the flow at a distance of 130 to 200 mm, i.e., 65 to 100 mesh diameters behind the grid. The mean velocity has been subtracted in order to highlight the turbulence structure. In obtaining the values for the flow map shown in Figure 1 the mean velocity in the flow direction (left to right in the picture) was first evaluated from the original grid of velocity components. The turbulence intensity, in the flow direction, was then evaluated by dividing the rms velocity fluctuation by the mean; in this case it was measured to be 6.2%.

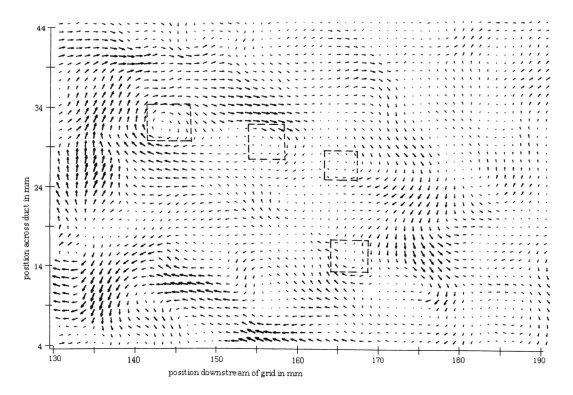

Figure 1 Velocity vector plot of grid-generated turbulence in a wind tunnel. Typical coherent structures are marked with boxes.

Even for a precise set of velocity vectors, significant errors can occur in the estimation of turbulence parameters due to the way in which statistical averages are evaluated. Computing mean and rms values from a single PIV velocity vector map is analogous to the determination of time-averaged mean and rms values from a point velocity record in a statistically stationary turbulent flow. In this case it is well known (Bendat and Piersol, 1968) that the mean square error (*MSE*) in measuring the time-averaged mean value of the velocity $U(t)$, i.e.,

$$\overline{U} = \frac{1}{T}\int_{t_0}^{t_0+T} U(t)dt \qquad (3)$$

is

$$MSE = \sigma^2 \qquad (4)$$

for short averaging times and

$$MSE = \frac{\sigma^2}{T}\int_{-\infty}^{\infty} K(\tau)d\tau \qquad (5)$$

for long averaging times, where T is the averaging time, $K(\tau)$ is the correlation coefficient (normalized autocovariance function) of the turbulence velocity record, and σ is the root mean square velocity fluctuation. It can be shown (Damm, 1995) that similar expressions exist for the spatial case. For a sample area of N1 × N2, the mean square error in the estimation of the velocity component in the flow direction is

$$MSE = \sigma^2 \qquad (6)$$

for small sample areas and

$$MSE = \frac{\sigma^2}{N1\,N2} \int_{-\infty}^{\infty} \int_{-\infty}^{\infty} K(\eta, \xi)\,d\xi\,d\eta \qquad (7)$$

for large sample areas (where correlation between velocity components at the extremities of the picture has been lost).

In order to evaluate this error for large sample areas in a particular flow situation it is necessary to assume a form for the correlation function. Taking a two-dimensional Gaussian distribution as being representative, i.e.,

$$K(\eta, \xi) = e^{-\alpha^2(\eta^2 + \xi^2)} \qquad (8)$$

we find that for large sample areas

$$MSE = \frac{\sigma^2 \pi}{\alpha^2\,N1\,N2} \qquad (9)$$

In deriving these expressions for the measurement error it is assumed that the flow is homogeneous over the region under consideration.

Coherent Structures

A great advantage of PIV, as compared to point measuring techniques, is the ease with which the results can be displayed and interpreted. Since velocities are obtained over a regular grid they can easily be converted into vorticity values by differentiation, i.e., if the velocity components are U and V in the X and Y directions, respectively, then, using the sign convention that a clockwise rotation corresponds to positive vorticity, the vorticity is given by dV/dX – dU/dY. A fifth-order polynomial is typically used for calculating the derivatives. In Figure 1, for example, four regions of particular structural interest have been marked by boxes. These can be more readily identified in Figure 2, which is the corresponding plot of vorticity. Note that even in grid turbulence definite structures are identifiable.

Vortex Wakes

An example of the strength of PIV flow mapping techniques in the identification of coherent structures is the study of wind turbine wakes at Edinburgh. Wind turbine performance is critically dependent on the geometry of the rotor wake, and the lack of detailed experimental data in the wake of a turbine and the difficulty of obtaining it are widely appreciated. Full-scale visualization experiments are difficult to perform and are limited by the problems of expense and nonrepeatable conditions. Qualitative descriptions of wind flow patterns and tip-vortex behavior were gained from flow-visualization studies (Pedersen and Antoniou, 1989) with the aid of smoke grenades attached to upstream masts or the trailing edges of the blades. The smoke studies confirmed the existence of a helical

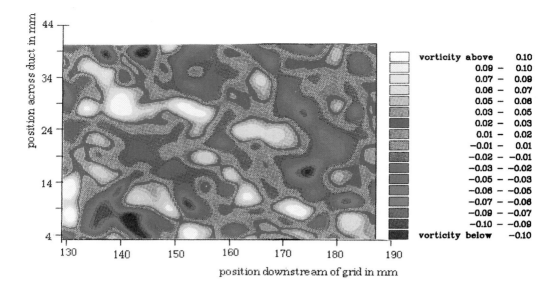

Figure 2 Vorticity map derived from the velocities displayed in Figure 1.

vortex system in the wake. The system comprises a weak, diffused vortex sheet core region which is shed from the trailing edge of the blades and assumes the form of a screw surface due to the rotation of the blades. In the near-wake region, this is dominated by an intense tip vortex helical system. For greater detail of the vortex structure, researchers have opted for wind tunnel testing using the techniques of hot wire anemometry (Vermeer and van Bussel, 1990) or laser anemometry (Green, 1985).

The use of PIV in the field of wind turbine aerodynamics is a recent development. Infield et al. (1994) have conducted tests both in wind tunnels and on a full-scale machine using pulsed lasers. The tests were restricted to the immediate vicinity of the blade and produced detailed profiles of bound circulation and the tip vortex. The study established the applicability and usefulness of the PIV technique as a velocimetry tool for wind turbines.

In order to visualize the full wake of a wind turbine rotor, PIV studies on small-scale model turbines have been carried out at the University of Edinburgh since 1991 (Whale and Anderson, 1993). The study has concentrated on capturing the near wake of a wind turbine. This region constitutes the most complicated part of the flow and numerical codes have particular trouble in modeling this region. A wide range of rotor operating states have been investigated. In particular the study has explored extreme operating states of the blades. These correspond to regions of flow where theoretical techniques give least satisfactory results. Flows in which the blade is heavily stalled are significant commercially to the wind turbine industry as current design methods for stall-regulated wind turbines are essentially empirical.

Wind Turbine Experiments

The experiments were carried out in a glass-based flume, equipped with a recirculating pump which allows a steady flow velocity to be established (Figure 3). The model turbine rig was placed across the tank, subjecting the rotor to a uniform current, U_o. Rotation of the rotor blades was achieved by driving the turbine with an electric motor/generator suspended on a frame above the water level. The ratio of the blade tip speed to the upstream flow velocity, known as the tip speed ratio, is

$$\lambda = \frac{\omega R}{U_o} \tag{10}$$

where ω is the angular velocity of the rotor and R is the rotor radius. The rotor is located at the end of an inverted tube, or tower, which was perpendicular to the oncoming flow to ensure symmetric inflow conditions. In order to reduce the disturbance to the rotor wake caused by the tower, it was streamlined with a plastic shroud of symmetric aerofoil cross section.

A two-blade model rotor of diameter 175 mm was used in the experiments. Flat-plate blades were chosen for their simplicity of construction and for reasons of comparison with theory, since the blades represent a very fundamental case. The blades have a thickness of 1.26 mm and a hub chord of 15 mm, with a linear taper to a tip chord of 10 mm. The turbine rig was placed in the flume in a position that captured cross sections of the helical vortex filaments shed from the trailing edges of the blades.

Conducting the PIV tests in water, rather than air, provided high seeding concentration and good illumination throughout the whole wake and overcame the problems of dispersion and condensation reported in wind tunnel tests. In addition, the lower speed of fluid and the higher speed of sound combined to avoid problems of compressibility effects encountered in high speed wind tunnels. An arrangement of honeycomb section, perforated plate, and fine mesh screen was used as flow manipulators upstream of the rotor. The manipulators aimed to reproduce turbulence length scales from values reported in field experiments. For these tests, the configuration was designed to produce uniform incoming flow and an upstream turbulence intensity of 4%.

Kinematic similarity between the model and a full-scale machine was achieved by running the model at tip speed ratios in the range λ = 2 to 8. These are tip speed ratios pertinent to the operation of a full-scale wind turbine. For the small-scale model, dynamic similarity with a full-scale turbine is less easily achieved. In wind turbine studies it is conventional to define a *blade* Reynolds number based on the velocity (W) and the chord (c) at a point on the blade. Thus

$$\text{Re} = \frac{Wc}{\nu} \tag{11}$$

where ν is the kinematic viscosity of the fluid. At a point on the blades located at 70% span, the blade Reynolds numbers of the model turbine can be calculated. Over the tip speed ratio range λ = 2 to 8, the Reynolds numbers fall in the range Re = 6.4×10^3 to 1.6×10^4. The Reynolds number of a full-scale wind turbine is typically of the order of 1×10^6.

In the PIV acquisition process, a Hasselblad 553 EL/X camera and 80-mm lens were used together with a rotating mirror image-shifting system. The shifting system was synchronized to photograph the two-bladed rotor in a vertical position, parallel to the tower of the rig. This captured cross-sections of trailing vortex filaments while avoiding interference of the blade in the sheet. In the synchronization process, the PIV recording captures the wake in the same phase. Averaging the instantaneous wake images extracts the coherent structure of the trailing vortex filaments from the superposed turbulence.

Results

Velocity measurements from the PIV tests for the two-blade flat plate rotor are presented in the form of vorticity contour plots, for tip speed ratios of 2.9 and 6.4. At the lower ratio, the rotor is in the windmill state (Figure 4). The wake is divided into regions of positive and negative trailing vorticity emanating from the tip of each blade. This is the expected

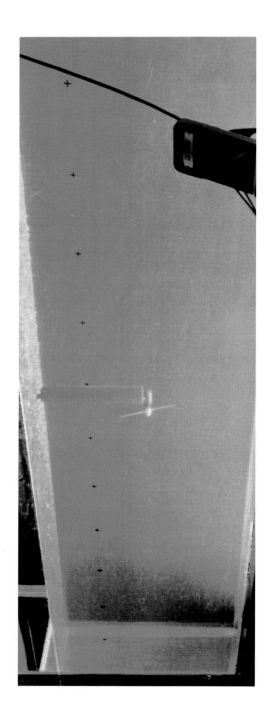

Figure 3 Two-blade model rotor in the water flume at Edinburgh.

Chapter seven: Analysis of turbulence and vortex structures by flow mapping 127

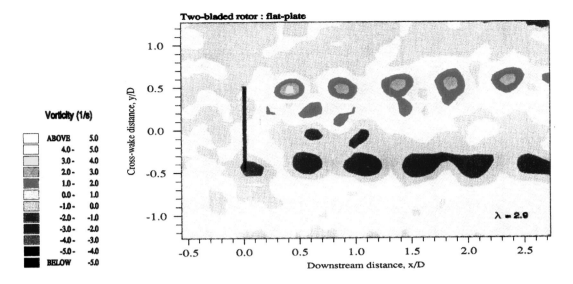

Figure 4 Vorticity map for the two-blade rotor at a tip speed ratio of 2.9.

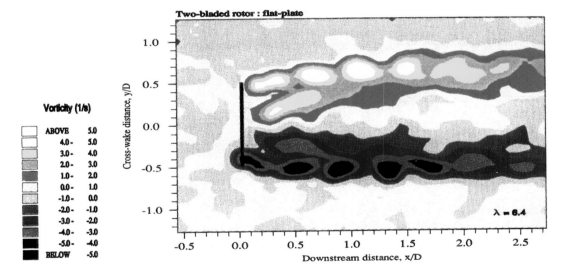

Figure 5 Vorticity map for the two-blade rotor at a tip speed ratio of 6.4.

pattern for the cross section of a helical vortex system. The shape of the boundary of the wake suggests mild wake expansion and concurs with the simple wind turbine wake theory (Wilson and Walker, 1974). At the higher tip speed ratio, the size and strength of the trailing vortices increase as the turbulent energy is preserved within the tip vortex (Figure 5). The wake is seen initially to expand, but then to contract at around one and a half diameters downstream. This has been observed to precede the breakdown of the structured wake into a highly turbulent state. The wake contraction is not considered in the standard wind turbine industry prediction codes, based on blade-element/momentum theory (BEMT).

Further examination of Figure 5 shows another major discrepancy with the simple wake theory. The rate at which the helical tip vortex structure convects downstream is seen to increase at about one diameter downstream, rather than continuing to decrease asymptotically. The velocity gradients across the inner part of the wake result in a shear layer of vorticity. The shear layer moves under the influence of wake expansion to merge with the tip vortex system. Simulations from current vortex wake codes do not display evidence of this inboard vorticity.

Conclusions

To date the principal merit of the PIV flow mapping technique has been in the identification of turbulent structures. Even in highly turbulent flows, coherent structures are usually present. These would normally be missed by point measuring probes, whereas they show up on PIV flow maps.

PIV flow maps can be used to gain quantitative information on turbulence statistics, but statistical errors are introduced analogous to those which arise with point measurements. CCD cameras are now becoming available which have good spatial resolution coupled with high frame rates. These will open up new possibilities for the study of turbulent flows where both spatial and temporal structures are required.

PIV has provided the means whereby, for the first time, detailed data from a complete instantaneous image of a wind turbine wake have been captured and recorded. Analysis of measured wake structures from the small-scale turbine may yield a fundamental understanding of rotor flows, in the same way that the principles of flight can be understood on the basis of model aircraft experiments. Although the Reynolds number of the PIV tests was lower than full scale by a factor of 1000, there is a strong possibility that the wake structures at different scales will share fundamental similarities.

The PIV results revealed properties of the wake structure which suggest that the current design codes used by the wind turbine industry are inadequate. PIV offers the potential to reveal the detailed physical processes governing wind turbine wake behavior. An advanced performance code that incorporates these phenomena will lead to the design of more efficient wind turbines.

References

Adrian, R.J. Optical methods for measuring vector velocity fields. I. Principles, *Lecture Series 1988–06,* von Karman Institute of Fluid Dynamics, 1988.

Batchelor, G.K. The theory of homogeneous turbulence, *Cambridge Science Classics,* 1986.

Bendat, J.S. and Piersol, A.G. *Measurement and Analysis of Random Data,* John Wiley & Sons, New York, 1968.

Damm, C.E. Particle image velocimetry, accuracy of the method with particular reference to turbulent flows, Ph.D. thesis, University of Edinburgh, 1955.

Durrani, T.S. and Greated, C.A. *Laser Systems in Flow Measurement,* Plenum Press, New York, 1977.

Gray, C. The development of particle image velocimetry for water wave studies, Ph.D. thesis, University of Edinburgh, 1989.

Gray, C. and Greated, C.A. The application of Particle Image Velicometry to the study of water waves, *Opt. Lasers Eng.,* 9, pp. 265–276, 1988.

Gray, C., Greated, C.A., McCluskey, D.R., and Easson, W.J. An analysis of the scanning beam PIV illumination system, *J. Phys. Measure. Sci. Technol.,* 2, pp. 717–24, 1991, reprinted in *Eng. Opti.*

Green, D.R.R. Modelling large wind turbine wakes, Ph.D. thesis, Loughborough University, 1985.

Infield, D., Grant, I., Smith, G., and Wang, X. Development of particle image velocimentry for rotor flow measurement, in *Proc. British Wind Energy Association 16,* Stirling, Scotland, 1994.

Jakobsen, M., Hossack, W.J., and Greated, C.A. Particle image velocimetry analysis using an optically addressed spatial light modulator: effects of nonlinear transfer functions, *Appl. Opt.,* 34(11), pp. 1757–1762, 1995.

Landreth, C. and Adrian, R.J. Electro-optical image shifting for Particle Image Velocimetry, *Appl. Opt.*, 27, pp. 4216–4220, 1988.

Meinhart, C.D., Prasad, A.K., and Adrian, R.J. Parallel digital processor system for Particle Image Velocimetry, in *Proc. 6th Int. Symp. on the Application of Laser Techniques to Fluid Mechanics*, Lisbon, Portugal, 1992.

Pedersen, T.F. and Antoniou, I. Visualization of flow through a stall-regulated wind turbine rotor, in *Proc. European Wind Energy Conf.*, Glasgow, 1989.

Skyner, D.J. The mechanics of extreme water waves, Ph.D. thesis, University of Edinburgh, 1992.

Vermeer, N.J. and van Bussel, G.J.W. Velocity measurements in the near wake of a model rotor and comparison with theoretical results, in *Proc. European Wind Energy Conf.*, Madrid, 1990.

Whale, J. and Anderson, C.G. An experimental investigation of wind turbine wakes using particle image velocimetry, in *Proc. European Wind Energy Conf.*, Lübeck-Travemünde, Germany, 1993.

Willert, C.E. and Gharib, M. Digital Particle Image Velocimetry, *Exp. Fluids*, 10, pp. 1811–193, 1993.

Wilson, R.E. and Walker, S.N. A fortran program for the determination of performance load and stability derivatives of windmills, Grant GI-41840, Department of Mechanical Engineering, Oregon State University, Corvallis, 1974.

chapter eight

Quantitative Visualization of 2-D and 3-D Flows Using a Color-Coded Particle Tracking Velocimetry*

Tzong-Shyan Wung

Institute of Applied Mechanics, National Taiwan University, Taipei, Taiwan, Republic of China

Abstract—*Modern digital imaging systems have extensive applications in flow visualizations because of their capability to display real-time field information with acceptable resolution. The recently developed color-coded particle tracking velocimetry uses these advantages to generate time-embedding streaks for low-speed flows. This paper summarizes the imaging and processing techniques for grabbing particle trajectories and revealing scenarios of particle movements in a slow-motion mode. The technique is also extended to acquire stereoscopic streaks for 3-D flows from two perspectives. Three components of velocity vectors can be directly deduced from stereopairs of time-coded streaks. Some quantitative visualization results of 2-D and 3-D flows are presented to demonstrate the capabilities of the method.*

Introduction

Due to rapid and progressional advancements in computer hardware and architecture, digital image systems have gained diverse applications in flow visualization in recent years. These applications include direct imaging of a fluid flow field in digital mode, and posterior image processing to extract quantitative flow information from the grabbed images or conventional photographic media. The use of digital image systems, indeed, largely facilitates developments of modern whole-field velocity measuring techniques.

Contemporary whole-field measurements are typically performed by a particle tracking velocimeter (PTV), or a particle image velocimeter (PIV) (Adrian, 1991). The PTV infers velocity vectors by measuring displacements of seeding particles from sequential exposures of the visualized flow. In a PIV, optical readout or digital transformation of double-exposed images in a small interrogating window provides displacements of particles, based on which a velocity field can be obtained. Both PTV and PIV usually illuminate the flow region with a sheet of light in order to determine planar velocity vectors. The velocity components perpendicular to the light-sheet cannot be deduced, which, if they exist, may increase the difficulty of data analysis for 2-D vectors.

---

* A version of this paper was presented at the *International Workshop on PIV*—Fukui'95.

When a 3-D velocity field is necessitated, the third component of a vector can be reconstructed from the velocity details of two or several adjacent planes (Cenedese and Paglialunga, 1989; Robinson and Rockwell, 1993). To investigate dynamic variations in unsteady flows, a laser sheet can be rapidly scanned over the flow region of interest to generate sequential 2-D images of seeding tracers which constitute a volumetric picture of the flow (Utami and Ueno, 1984; Nosenchuck and Lynch, 1986). An alternative to sweeping of a light-sheet is to illuminate the fluid volume with multiple colored light sheets. Scattering from multiple colors are simultaneously collected by multiple cameras with corresponding color filters (Mantzaras et al., 1988).

Those light-sheet-based techniques essentially provide a set of flow information on the planes of light illumination which are appropriate for flows with dominant velocity components along the planes. The third velocity components of the flow field are derived in an indirect way. To obtain a 3-D picture of the flow, stereoscopic imaging is a conventional approach for acquisition of three components of velocity vectors. Images of seeding particles are produced simultaneously from two different angles on a pair of planar recording media such as a photographic film or a CCD (charge-coupled device) sensing array. Positions of particles of multiple exposures or multiple frames are processed to recover velocity information and path lines. Recently, PIV has been extended for measurements of three-dimensional vectors (Gauthier and Riethmuller, 1988; Prsad and Adrian, 1993). In the 3-D PIV the thickness of the illumination volume is maintained in a thin layer to avoid possible complexity and ambiguity of matching stereo particle image pairs.

A novel color-coded particle tracking velocimetry (CCPTV) has recently been introduced and implemented by Wung and Tseng (1992) for visualization of a 2-D flow field and generation of colored particle streaks with embedded temporal marks in real time. This paper outlines the image grabbing technique for temporal mark encoding, then follows by the bit-plane image operation for particle movement observation. Next, a posterior image processing procedure for velocity extraction is described. The technique is further extended to form a stereo imaging system for direct realization of 3-D velocity vectors. Stereoscopic trajectories of particles are actually acquired by one imaging system through a two-view-producing device. Some quantitative visualization results will be presented, including natural convection in an enclosure, lid-driven cavity flow, and 3-D flow in a square duct and in a cube with offset inlet and outlet. Since this paper emphasizes the quantitative visualization technique and demonstrates visualized results, detailed physics of the flows will not be discussed here.

Image Grabbing and Coding

The present CCPTV utilizes a digital imaging system for quantitative visualization of low-speed flows. Figure 1 shows the flowchart for the image acquisition. Flow situation is imaged by a CCD camera which sends the video signal to the image processing system at a scanning rate of 30 frames per second for the NTSC RS-170 format. This analog video signal is then digitized into a digital image frame with a resolution of 512×480 pixels. Each pixel has an 8-bit data which can represent 256 (2^8) possible gray levels. Those gray values conventionally portray brightness of seeding particles in the fluid as they are exposed to the illuminating light.

The innovation of the CCPTV is the utilization of gray scales of a pixel to store both the timing knowledge and the particle trace required for deriving the particle velocity (Wung and Tseng, 1992). In order to resolve flow information for each time duration, an 8-bit image frame is allocated as eight bit-planes into which trajectories of particles of a time period are sequentially and independently stored (Wung et al., 1994). To do this, the digitized images carrying flow details are conducted to an input look-up table (ILUT) for binarization and temporal marks embedding. A look-up table (LUT) is a hardware

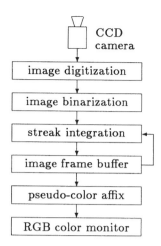

Figure 1 Image processing flowchart.

memory block consisting of 256 entries with 8 bits each. An input pixel value to the LUT is used as an address, and the value stored at that address is output as the transformed pixel value. By predefining the LUT values, any pixel transforms can be performed. In the present visualization technique, the ILUT is programmed in the way that it binarizes the input live image with the output values steppedly varying with time intervals to bear temporal code. In practice, during the first time interval, the output gray scale is set to 1 if the input pixel value is larger than a predetermined threshold value which signals a passing particle. Otherwise, the output value is replaced by 0.

The use of a threshold value for image binarization acts as a high-pass filter which can filtrate out most noises and selects an appropriate number of light-reflecting particles for velocity extraction. Determination of the threshold value depends on experimental conditions such as intensities of luminescence, light scattering of particles, background noise, and so forth. A few trial-and-error tests will provide a proper quantity for the binarization.

In the second time interval, the output value of the binarized image is set to 2, equivalent to the second bit value. In the next following time intervals, the output gray levels are sequentially set to 4, 8, 16, 32, 64, and 128, corresponding, respectively, to the binary values of the third to the eighth bit-planes. By doing so, the timing information can thus easily be implemented as gray levels during image grabbing. A nonzero value in a certain bit of a pixel simply dictates that a seeding particle has appeared at the position where the sensing pixel is mapped during that corresponding time interval.

The images displayed after binarization contain discrete moving particles with their gray levels changing with time. These digital images are generated at a rate of 7.37 megabytes per second which is beyond the capability of most personal computer-based image systems for data transfer and storage over an extended time period. Also, those discrete images of particles are not suitable for velocity determination. The CCPTV, however, effectively integrates these images of particles into color-coded streaks in real time using low cost equipment.

The streak integration is carried out by employing a real-time arithmetic-logic unit (ALU) in the image system. The ALU manipulates two input pixel values, one from the binarized live image and one from image frame buffer, then outputs the results to the image frame buffer. The frame buffer stores the accumulated image and sends the stored pixel values serving as feedback signal. Both the video camera and the image processing system are synchronously scanned with the same starting address so that pixel values of the same address respectively feed from the live image and the frame buffer arrives at the

ALU at a same clock pulse. The ALU executes a logical *OR* operation on the two input values and outputs the result to the frame buffer replacing the value in the same address. This operation results in integration of particle streaks to the corresponding bit-plane of an image frame over a time period. By steppedly changing the outputs of the image binarization to the sequential bit values for each time interval, trajectories of particles are properly grabbed and encoded with temporal marks.

The length of a time interval can be an arbitrary multiple of 1/30 s for the RS-170 video format with the shortest duration for one trajectory segment being one framing time (i.e., 0.0333 s). By the present image processing, a sequence of video images in a time period is likely to be compressed and stored in a bit-plane in real time. Also, trajectories of eight time intervals are separately, independently, and integrally recorded in eight bit-planes. Therefore, the CCPTV effectively utilizes the digital data structure for image storage which reduces the required memory space of the system.

As the digital system proceeds image acquisition, temporal mark encoding, and trajectory accumulation, image stored in the frame buffer is routed to output look-up tables which affix pseudocolors to the streaks according to the gray-scale values of the image. Then, the pseudocolored streaks are simultaneously displayed on a color monitor. The colored streak images not only largely enhance human visual perception in observation of particle traces, but also explicitly dictate moving directions of seeding particles by the color code.

Posterior Image Processing

Because of independent storage of particle streaks over a time interval in a bit-plane, several particular binary image processes can be quickly performed on the acquired flow information. For example, particle streaks of any time interval can be easily singled out for revealing or further processing by simply using a pixel mask consisting of only a bit value of 1 at a desired bit-plane and logical *AND* operating with each pixel of the image frame, as shown in Figure 2(a). The resultant image is then the particle tracks of that particular time interval, while trajectories of the other bit-planes are totally removed.

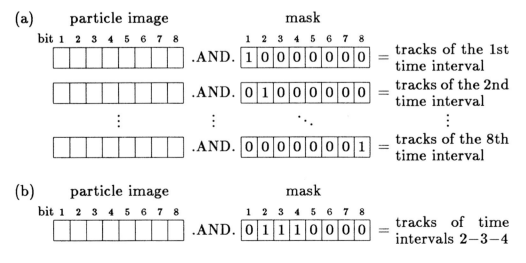

Figure 2 Manipulations on the bit-plane images. (a) Sequential display of particle tracks of eight time intervals; (b) display of streaks of three contiguous time intervals. (From Wung, T.-S., Shyu, K.-S., and Lee, S.-W., 1994, *J. Chin. Soc. Mech. Eng.*, Vol. 15, No. 5, pp. 431–442.)

As the binary value of the pixel mask sequentially shifts its position and logical *AND*s with each pixel of the image frame as shown in Figure 2(a), the grabbed streaks can be displayed plane-by-plane so that the scenarios of particle movements are played back in a slow-motion picture mode. The displaying time for each bit-plane can be arbitrarily set in proportion to the image integrating duration. Reveal of the flow motion can be carried out immediately after the images have been captured, allowing one to decide whether the results of visualization require further processing or not.

If three contiguous bits of the pixel mask are set to 1 and the rest of the bits to 0, as sketched in Figure 2(b), the *AND* manipulated image using this mask will contain particle traces of three bit-planes. Any segment of a trajectory having coded traces on its two ends is eligible for velocity determination. The length of the middle segment provides the velocity vector, and the gray levels of streaks indicate sense of the vector. This three-bit mask filtrates out irrelevant streaks and simplifies posterior image processing for velocity quantification. It is obvious that six velocity vector fields can be derived from an 8-bit image frame in which particle tracks of eight time intervals are stored.

Due to the architectural structure of a CCD camera and operation format of a video signal, streaks captured by the CCPTV often consist of fragmental segments. These fragmentations can be attributed to interlaced video scanning mode, photon-excited electronic charge transfer in a CCD chip, interruption of image processing when waiting for vertical synchronous signal, and three-dimensionality of the fluid flow field (Wung, 1995).

To facilitate automatic extraction of velocity vectors from the time-encoded streaks, Wung and Lee (1994) employed a morphological image processing to fill up interline discontinuities and gaps up to two pixels. Then, particle tracks are skeletonized by medial axis transformation for multiple gray-scaled images. Thus, velocity vectors can be easily quantified by tracing the thinned paths and the associated time code. This image processing procedure is illustrated in Figure 3. Figure 3(a) portrays the pseudostreaks of a 2-mm LED with embedded time codes. The interline discontinuities can be clearly seen on the two ends of each segment. Those fragments were filled up by morphological dilation in Figure 3(b) and by erosion in Figure 3(c). The dilation can be performed more than once, depending on the allowable free space for the image enlargement to avoid interference

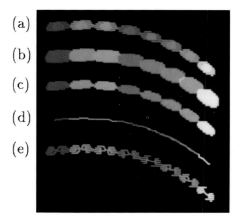

Figure 3 Morphological processing for the time-coded streaks. (a) The original track images consisted of interlaced discontinuities; (b) enlarge the track images by dilation operation; (c) reduce the track images to their original widths by erosion operation; (d) one-pixel-width skeleton by the medial axis transformation; (e) irregular skeletons by direct skeletonization on the raw image without closing.

with the neighboring objects as well as the sizes of holes and gaps to be filled. The expanded track shown in Figure 3(b) was dilated twice by using a 3×3 contiguous structuring mask, although dilating once could effectively remove the interline discontinuities. The expanded pixels in a dilated image could be removed by the morphological counterpart of dilation transformation. Eroding a dilated image by the same structuring element would delete the swelled pixels along the outer boundary of the image. The number of erosion operations on a dilated image should be balanced with that of dilations to avoid distorting the geometrical representation of the track image. A comparison between Figures 3(a) and 3(c) clearly evidences that the interline blanks have been closed, and that the boundary contours of the tracks remain unaltered.

Next, the closed streak is skeletonized to a pixel width by the medial axis transformations in Figure 3(d). The velocity vectors can be easily calculated by tracing the one-pixel-width track and the associated time code. Figure 3(e) shows the strange result of direct skeletonization of the raw image without closing. It is obvious that the skeletons faithfully reflect the interlaced images in the extremes of each segment of the time-coded tracks. The original interlaced image may actually hinder the process for velocity determination, except for the very slow fluid motion where the interlaced blanking is not pronounced.

2-D Visualization Results

Figure 4 demonstrates qualitative and quantitative visualizations and velocity vector field of natural convection in a rectangular enclosure with heating from bellow and cooling on the top wall. The enclosure had a cross-sectional area of 40×40 mm^2 and a length of 380 mm. The enclosure was filled with glycerin as the working fluid and seeded with aluminum powder of size between 20 to 40 μm. Figure 4(a) displays a qualitative visualization of the convection when the enclosure inclined 15° about its longitudinal axis at a Rayleigh number of 5×10^4, based on the height of the enclosure and the temperature difference between the heating and cooling walls. In this picture, trajectories of seeding particles were accumulated using an OR operation over the input video images. This realization without embedding temporal marks is equivalent to long-time exposure of a photograph to the flow with the advantage of obviating overexposure of the film. The time-embedded streaks shown in Figure 4(b) consist of four durations, consequently colored as green, yellow, red, and blue. The color code clearly dictates flow directions. The quantified velocity vectors shown in Figure 4(c) are average quantities over the two inner segments, although they can provide two sets of velocity vectors. With properly seeded particles, the flow field can be quantitatively visualized capable of providing velocity field on the light-illuminating sheet. It was found from visualizations of several cross sections that, as the enclosure is included at 15°, a longitudinal rolling cell virtually fills the enclosure which can be regarded as 2-D flow.

Figure 5(a) shows a quantitative visualization of a lid-driven cavity flow for a Reynolds number of 600, based on the lid velocity and the height of the cavity. The cavity having a square cross-sectional area of 10×10 mm^2 and a depth of 200 mm was filled with water as the working fluid. A belt in tension was moved from right to left on the top as the lid of the cavity. The visualized section was on the symmetric plane of the longitudinal direction. Eight colored segments were recorded in an image frame. It can be seen that the moving lid causes the fluid in the cavity to circulate about the center slightly downstream. Several wide traces in the figure may be attributed to contaminants or oxides of aluminum powder. Figure 5(b) plots contours of computed 2-D stream functions. A comparison between the visualization and computation reveals good agreement, although the experiment did not provide streaks in some low speed regions.

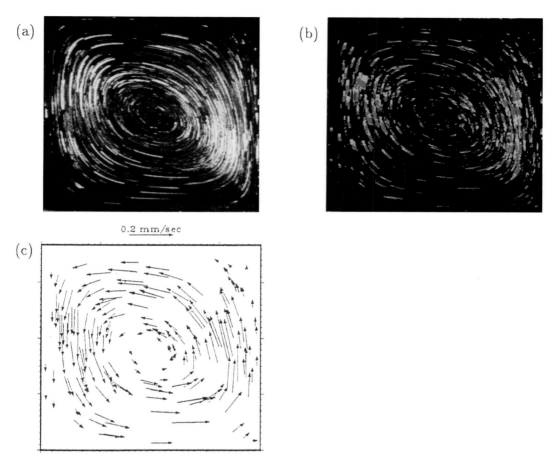

Figure 4 Visualization of natural convection in a rectangular enclosure for Ra = 5 × 10⁴ and inclination angle 15°. (a) Qualitative visualization; (b) time-coded streaks for the flow field; (c) quantified velocity vector field from (b).

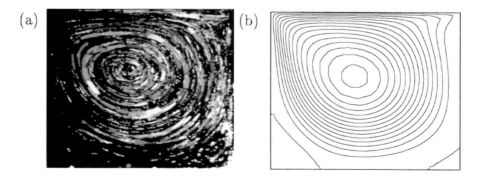

Figure 5 Visualization for the lid-driven cavity flow at Re = 600. (a) Time-coded streaks; (b) contours of computed stream function for the flow.

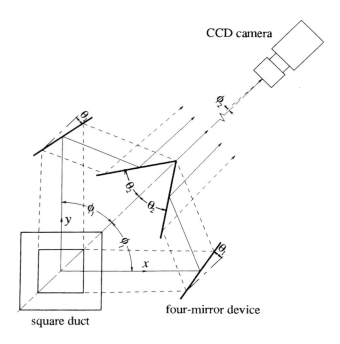

Figure 6 Arrangements for stereoscopic imaging by a single camera through a four-front-mirror device.

Stereo Imaging and 3-D Results

The above described imaging technique is also suitable for quantitative visualization of 3-D flows through stereoscopic methods. Conventional stereoscopic approaches observe an object simultaneously from two slightly different angles. The stereo color-coded particle tracking velocimetry (SCPTV), however, visualizes a flow field from two orthogonal directions. Figure 6 shows the schematic of the experimental setup for stereoscopic realization of fluid flow in a square duct or in a cube. Instead of using two imaging systems for stereoscopic purpose, a four-front-mirror device is employed to bring two-view images of the flow to a single CCD camera for image grabbing. Such an arrangement requires only one imaging system with a cost of reducing the spatial resolution to half.

The four-mirror device was used for stereophotography shortly after the invention of photography in the mid-19th century (Okoshi, 1976). It is also a convenient tool for survey of photogrammetic image pairs in a stereoscopic mode. Kent and Eaton (1982) employed the device incorporated with a high-speed cinecamera to investigate flows in an engine cylinder. In the device, four plane mirrors are arranged as two symmetric pairs of double reflectors which conduct two perspective images into the focus lens of the camera, and then project onto the sensing area as two adjacent halves. The two mirrors in each channel are positioned at θ_1 and θ_2 with respect to the symmetric line as shown in Figure 6. These angles are related to the angle of parallax of the two views $2\phi_1$ and the angle of field of view ϕ_1 for each channel by

$$\theta_2 - \theta_1 = \frac{1}{2}(\phi_1 + \phi_2)$$

It is obvious that there are many possible arrangements satisfying the above condition. To select a proper layout of the mirrors, the first consideration must avoid obstruction of reflected images, and the second must avoid a large angle of incidence which is the angle

Chapter eight: Quantitative visualization of 2-D and 3-D flows   139

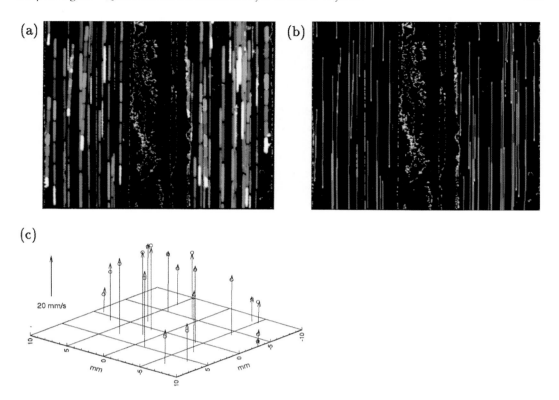

Figure 7   Stereoscopic flow visualization in a square duct at Re = 388. (a) Color-coded streaks; (b) thinned trajectories of tracers; (c) distribution of velocity vectors. Arrowhead: experiment; circle: computation.

between the incident ray and the normal to the mirror at the incident point. The symmetric configuration will give the same length of optical paths for both channels so that they have the same magnification scale for the recorded stereo images. Detailed layout of the four mirrors can be found in Wung and Lu (1995).

The two-view images are displayed side-by-side vertically on a monitor in which each view has a resolution of 256 × 480 pixels. There exists, however, an overlapped region of several pixels between the two images of a practical stereoscopic picture mainly due to the knife edge effect when a mirror is placed in-between the lens and the object, as well as the finite object distance from the lens. Consequently, the effective recording area for each view reduces to about 200 × 480 pixels.

Unlike conventional PTV or PIV techniques using a sheet of light, seeding particles in the fluid are illuminated by a projection light to cover the entire measuring volume. For the first stereo visualization, the square duct has a cross-sectional area of 20 × 20 mm^2. The test section is located 350 mm downstream of the inlet of the square duct. Images of seeding tracers are simultaneously scanned from two perpendicular views by a single imaging system through the four-mirror device. Trajectories of particles are accumulated and coded according to the algorithm of the CCPTV in the two perspectives as a whole picture frame. The stereo pairs of trajectories are then matched to find the actual positions of the particles, from which three components of velocity vectors can be directly resolved.

Figure 7(a) demonstrates stereoscopic color-coded particle streaks of a fully developed square duct flow for a Reynolds number of 388, based on the hydraulic diameter and mean flow velocity in the duct. Water with pliolite particles of size between 150 to 200 μm flowed downward in the picture. The square duct was fabricated by 8-mm plexiglass plates for ease of light illumination and image acquisition. Some irrelevant areas such as walls or contaminant spots have been removed since they displayed a brightest gray level of 255. In this realization, the left half is the right-hand-side view of Figure 6, and the right half the top view, with the x- and y-axes on a cross section of the square duct and the z-axis along the axial direction.

A glance of the figure may not be easy to match stereoscopic pairs of particle streaks, because some trajectories of a time interval were overlaid by the others. Also, streaks of the whole observing volume were projected onto a planar sensor which records no depth information. As described above, if one isolates streak images of a bit-plane from an 8-bit image frame, streaks of the other time intervals are filtrated out. Then, the density of displayed streaks reduces to about 1/8 of the original image, thereby largely facilitating accessibility of posterior image processing. By sequential displaying bit-plane images, scenarios of particle movements in the two perspectives can be revealed.

Figure 7(b) shows the thinned trajectories of Figure 7(a) as a result of the above-described morphological image processes and skeletonization. Since streaks of several time intervals might be overlapped, the posterior image processes were performed on each bit-plane, and then resumed to its original bit-plane which affixed with a false color. Once the trajectories can be represented by their skeletons, they are ready to match and pair in order to find 3-D velocity components.

Stereoscopic pairing of two-view images of tracers can be difficult, especially for a large angle of parallax, if the seeding tracer density is not low. Fortunately, color-coded streaks contain more information for determination of 3-D positions of tracers that are extreme positions of streaks and the associated gray levels. As a terminal end of a streak represented by a pixel, a light ray in the space can be defined which projects onto the sensing element through the lens. An equation of light ray in the fluid layer can be derived simply employing the ray tracing technique to follow a light through the observing window layer (Wung and Lu, 1995). The actual position of the particle must be determined by the image of the second view. An equation of the light ray for the represented pixel of the second view can be similarly established. The intersection of the two light rays gives the location of the tracer in the flow channel. As analysis by Wung and Lu (1995), the image shift due to light refraction through layers of window and fluid can reach up to 7 pixels for the present square duct setup. This refractive displacement not only affects stereo pair matching, but also decreases the accuracy of velocity determination. Therefore, correction on the streak image is necessary, especially for the case of using one imaging system to grab stereo images.

Figure 7(c) plots velocity vectors quantified from Figure 7(b). There are 19 matched stereo trajectory pairs in Figure 7(b) from which 60 valid 3-D velocity vectors can be obtained. Since the test section is situated in the fully developed region, all velocity vectors are brought to a reference section so that the traces can be clearly located with the flow direction pointing upward. Seeding tracers are randomly populated in the duct flow. To make a quantitative comparison, Figure 7(c) also marks circles which indicate the computed velocities at the corresponding tracer locations. It can be seen that the experimental vectors agree very well with the computed results. The accuracy is within 7.4% when the differences are normalized by the mean flow velocity in the duct. This value is regarded acceptable for the present real-time image grabbing, whole-volume flow visualization technique.

It is noted that any streak segment with proper temporal marks on both ends is eligible for velocity determination, not necessarily consisting of all eight colors. To determine a 3-D velocity vector, a stereo streak pair is necessary. However, some particle tracks recorded on one view many not appear on the other, which may attribute to nonspherical shapes of the tracers, nonuniform illumination, high threshold value, etc. Those unmatched trajectories are rejected during data reduction for three velocity components, except for straight trajectories in which the stereo pairing conditions in one view can be relaxed.

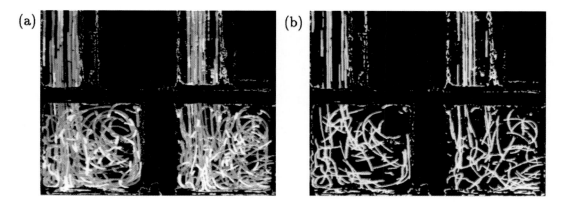

Figure 8 Stereoscopic flow visualization in a cube at an inlet Reynolds number of 948. (a) Color-coded streaks; (b) streaks of segment 3, 4, and 5, respectively, colored by blue, yellow, and cyanine, overlapping with skeletons on white.

The above fluid flow case has a dominant velocity component along the longitudinal direction of duct. The main objective of the experiment is to validate stereo imaging and streak pairing processes. The next realization is a square duct flow passing through a cube with the inlet and outlet arranged on the diagonal. The above-described fully developed duct flow serves as the inlet at a lower corner of the cube. The cube has an inner dimension of $50 \times 50 \times 50$ mm^3, with an outlet duct of 20×20 mm^2 having a diagonal offset and a 90° turn with respect to the inlet.

As the fluid enters the cube, the main stream has to change its flow direction in order to reach the outlet, hence inducing circulation regions in the cube. Qualitative visualization using color dyes shows that the flow in the cube is 3-D with unsteady eddies revolving in the chamber (Wung and Teng, 1996). Figure 8(a) shows a quantitative visualization of the cube flow for an inlet Reynolds number of 948. This two-view image is arranged similar to Figure 7(a) with a larger object distance. It seems very difficult to match and to pair trajectories of tracers from this stereoscopic photograph. To take advantage of the color-coded streaks, Figure 8(b) portrays streak images of three contiguous bit-planes, the third, fourth, and fifth time intervals, respectively, colored by blue, yellow, and cyanine. The skeletons of streaks, colored in white, are also overlaid in the photograph. The flow in the cube involves strong 3-D motion as evident by the visualized streaks. Investigation of the flow physics from visualization results requires identification of flow regime in the cube incorporated with qualitative observations. Details of the flow physics will be documented in a subsequent paper (Wung and Teng, 1996).

Concluding Remarks

The principle and image processing of the color-coded particle tracking velocimeter is summarized in this paper. The technique is extended to form a stereoscopic system through a four-front-mirror device. In 2-D or 3-D visualization, images of seeding particles are recorded and encoded with temporal marks in real time. Those quantitative realizations can be easily processed to provide 2-D and 3-D velocity information.

The accuracy of the quantified velocity depends on the length of a valid segment for vector calculation. The longer the length of a segment, the more accurate velocity will be obtained. Because of limited spatial resolution and video scanning rate of the digital imaging system, the present technique employing whole-volume illumination permits a low density of seeding particles in order to reduce difficulties in the posterior processing for velocity vector extraction. For steady flows, one can collect trajectories of particles over several image frames to construct a complete flow picture. For periodic unsteady flows, one may synchronize the imaging system with the oscillating phase of the flow so that image accumulation over periods is feasible.

Nevertheless, the CCPTV provides an alternative and effective tool for the real-time realization of the flow using inexpensive equipment with less labor. The SCPTV demonstrates its capability of simultaneous and direct measurements of three velocity components in a fluid flow field.

Acknowledgment

This work was supported by the National Science Council, Republic of China, under grant number NSC 84-2212-E-002-022.

References

Adrian, R. J., 1991, "Particle-Image Techniques for Experimental Fluid Mechanics," *Ann. Rev. Fluid Mech.*, Vol. 23, pp. 261–304.
Cenedese, L. and Paglialunga, A., 1989, "A New Technique for the Determination of the Third Velocity Component with PIV," *Exp. Fluids*, Vol. 8, pp. 228–230.
Gauthier, V. and Riethmuller, M. L., March 1988, "Applications of PIDV to Complex Flows: Measurement of the Third Component," *VKI Lectures Series on Particle Image Displacement Velocimetry*, Brussels.
Kent, J. C. and Eaton, A. R., 1982, "Stereophotography of Neutral Density He-Filled Bubbles for 3-D Fluid Motion Studies in an Engine Cylinder," *Appl. Optics*, Vol. 21, pp. 904–912.
Mantzaras, J., Felton, P. G., and Bracco, F. V., 1988, "Three-dimensional Visualization of Premixed-charge Engine Flames: Islands of Reactants, and Products; Fractal Dimensions; and Homogeneity," SAE/SP-88/759 Proc. of the Int. Fuels and Lubricants Meeting and Exposition, Portand.
Nosenchuck, D. M. and Lynch, M. K., 1986, "Three-dimensional Flow Visualization Using Laser-Sheet Scanning," AGARD-CPP-413, Paper #18.
Okoshi, T., 1976, *Three-Dimensional Imaging Techniques*, Academic Press, New York.
Prasad, A. K. and Adrian, R. J., 1993, "Stereoscopic Particle Image Velocimentry Applied to Liquid Flows," *Exp. Fluids*, Vol. 15, pp 49–60.
Robinson, O. and Rockwell, D., 1993, "Construction of Three-Dimensional Images of Flow Structure via Particle Tracking Technique," *Exp. Fluids*, Vol. 14, pp. 257–270.
Utami, T. and Ueno, T., 1984, "Visualization and Picture Processing of Turbulent Flow," *Exp. Fluids*, Vol. 2, p. 25.
Wung, T.-S., 1995, "On the Video Image of Particle Streaks", *Bulletin of the College of Engineering, National Taiwan University*, No. 63, pp. 47–66.
Wung, T.-S. and Lee, S.-W., 1994, "Quantitative Visualization and Image Processing for Slowly Unsteady Flows," *Proc. 3rd Asian Symp. on Visualization*, Japan Society of Visualization, Tokyo, pp. 651–656.
Wung, T.-S. and Lu, J.-Y., 1995, "Three-dimensional Positioning of a Particle from Stereo Pictures Imaging by One Video Camera," *J. Chin. Soc. Mech. Eng.*, Vol. 16, No. 2, pp. 141–153.
Wung, T.-S., Syu, K.-S., and Lee, S.-W., 1994, "A Quantitative Visualization Technique for Slowly Unsteady Flows," *J. Chin. Soc. Mech. Eng.*, Vol. 15, No. 5, pp. 431–442.
Wung, T.-S. and Teng, K.-Y., 1996, "Investigation of 3-D Flow Fields in a Cubic Chamber Using a Stereo Color-Coded Particle Tracking Velocimetry," to appear.
Wung, T.-S. and Tseng, F.-G., 1992, "A color-Coded Particle Tracking Velocimeter with Application to Natural Convection," *Exp. Fluid*, Vol. 13, pp. 217–223.

chapter nine

Visualization and Determination of Local Mass Transfer at Permeable and Nonpermeable Walls in Liquid Flow

W. Kühnel and V. Kottke

Institut für Lebensmitteltechnologie, Universität Hohenheim, Stuttgart, Germany

> **Abstract**—The local heat and mass transfer in liquid flow at solid walls is visualized and determined by an experimental method based on chemisorption of a dye. The surface of the sample is coated with a thin layer of a chemisorbent, to which the dye is chemisorbed resulting in a color intensity distribution. This distribution corresponds directly to the local mass transfer. The color intensity distribution is measured by remission photometry or alternatively by digital image processing. With a calibration equation, the measured remission is converted to the local surface mass density. With the flow parameters the local mass transfer coefficient can be calculated. The diffusion coefficient is measured in a diaphragm cell, thus the Schmidt number and the local Sherwood number can be determined. The visualization technique is applied in investigations of multiple jet-cleaning systems by impinging jets, sinusoidal wavy channels, spacers in membrane systems, and plate heat exchangers.

Introduction

In many processes with, for example, heat exchangers, regenerators, membrane systems, catalysts, adsorbers, column inserts, and static mixers, the local and integral heat and mass transfer is very important for the effective use of energy and material. Therefore, for the optimization of these processes the major aim is the investigation and the enhancement of the heat and mass transfer. By using methods for visualization and quantification — here especially the wall tracing, the visualization of the local mass transfer at solid walls — it is possible to visualize the flow even in highly complex turbulent flow fields. Besides a number of approaches to calculate flow fields together with progresses in numerical computation of turbulent flow, the numerical determination of the local heat and mass transfer is not possible in many cases.

A visualization technique which enables the quantification of the integral as well as the local heat and mass transfer is very useful. In previous publications (Kottke et al., 1977) a technique for gas flow was presented even for complex geometries. This technique can be applied in experimental devices as well as in plants in industrial scale.

In many cases it is not allowed to transfer the results from experiments in gas flow to liquid flow, e.g., separated flow, internal flow, gas–liquid flow, flow in membrane modules, and non-Newtonian media. Therefore, a new technique for the visualization and quantification of mass transfer in liquid flow has to be developed.

Visualization in Liquid Flow

Experimental methods for the visualization and determination of heat and mass transfer in liquid flow can be distinguished in three basic types: electrochemical methods, solution techniques, and chemisorption techniques. Methods based on chemisorption enable the visualization as well as the quantification of the local and the integral mass transfer, while the experimental expenditure is rather small.

Chemisorption methods can be used to visualize the local mass transfer with a high local resolution. The boundary layer flow is not influenced by this technique, and by using optical measuring techniques, the visualized mass transfer can be quantified. A dye is dissolved in the liquid and the sample is coated with a chemisorbent. During the experimental run the dye is fixed to the chemisorbent and the amount of dye locally fixed corresponds directly to the local mass transfer. The result is a color intensity distribution. An area with dark color intensity corresponds to a high mass transfer and vice versa. Depending on the system dye/coating, the visualization has a high local accuracy and resolution. The fixation is permanently stable.

The New Technique

The technique presented here for the visualization and determination of heat and mass transfer in liquid flow is based on chemisorption of a dye at solid surfaces.

In a simple experimental apparatus (Kottke et al., 1992) for the visualization of an impinging jet, different systems of dyes and coating materials were tested. The quality of the visualization was valued by examination of the visualized stagnation point. The criteria for the quality are the stability of the visualization and the accuracy and local resolution. Many dyes showed bad results. They were not permanently fixed. Thus the visualization was very weak. These dyes were not fixed by chemisorption because they could diffuse onto the wet surface. Therefore a fundamental condition is that each dye molecule reaching the surface must be fixed immediately to the surface coating.

The best results were achieved with a blue anionic dye (Acidol Blau 3GX-N 200%; BASF Germany) which has a high affinity to polyamide. The dye is combined with a chromatographic foil coated with a thin layer of polyamide, polyamide membranes, or polyamide foils. These foils or membranes are mounted onto the surface of the sample which has to be investigated. The dye is dissolved in the liquid. By a salt bond — electrostatic binding — the dye is permanently fixed to the polyamide of foil or membrane.

This fixation by chemisorption is very important as can be observed in the experiments with the apparatus for stagnation flow. The single dye molecule must not diffuse on the surface, otherwise the visualization of the locally transferred mass cannot be realized quantitatively. If the dye is not fixed by chemisorption the molecule is moving as long as the surface is wet, and the resulting color intensity distribution would change while the sample is removed from the experimental apparatus. Then a quantification produces wrong results.

Figure 1 shows an example of a visualization of the stagnation point of an impinging single jet. The nozzle was submerged and the impinging jet had a Reynolds number of $Re = 27,000$. The dimensionless distance between nozzle orifice and the wall was $h/d = 1.5$. The visualization agrees very well with the investigations of Martin (1977) and Carlomagno (1992). It shows the mass transfer in two major concentric maxima.

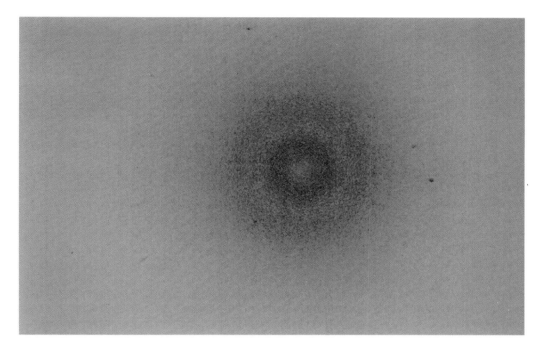

Figure 1 Mass transfer distribution as color intensity distribution in an impinging jet flow (Acidol–Blau and polyamide).

Optical Measurements

The local mass transfer is determined by optical measurements. After the experimental run the foil is removed from the apparatus and dried at ambient temperature. The drying of the foil is essential to avoid the influence of the wetness on the color intensity. After a certain time of drying (1 h), which was experimentally determined, the mass transfer can be measured.

Two different optical measuring systems are applied: remission photometry and digital image processing.

Remission Photometry

The remission photometry is well established (Kühnel and Kottke, 1994). Here the measured remission corresponds to the color intensity, which depends on the locally transferred mass. The remission is measured along several defined lines by using a simultaneous spectral photometer in combination with a cross table. The number of lines depends on the complexity of the visualized structure. The whole system is fully automated and controlled by a computer. The measured remission is transformed by applying a calibration equation to the local surface mass density and the local mass transfer coefficient. The time for the evaluation of one sample depends on the sample size and the number of lines.

Digital Image Processing

A new system for measuring the color intensity distribution is a digital image processing system which allows the evaluation of the entire sample in a few seconds. The system is based on a black and white slow scan CCD camera with a resolution of 14 bits and a pixel

Digital Image Processing

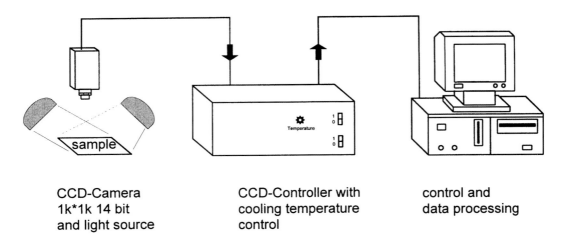

Figure 2 Digital image processing system for measuring color intensity distributions.

array of 1024 × 1024 pixels. The camera is cooled and controlled by the control unit. A personal computer controls the whole system and processes the data. The light source is an essential part of the measuring system besides the CCD camera. Here various different light systems were tested, and as a result of these tests it is necessary to use a stabilized light source with a diffuse light distribution. Figure 2 shows the setup of the digital image measuring system.

The locally measured gray scale depends on the local color intensity and the local light intensity. Thus it is necessary either to illuminate the sample completely homogeneous or to compensate the inhomogeneous lighting by an appropriate algorithm, the so-called shading correction.

Figure 3 shows the scheme of how the local mass transfer coefficient is determined from a colored sample by using an experimental visualization method based on convective mass transfer. The resulting picture is the product of three single pictures: the background noise $\overline{\Delta}$, the reference picture $\overline{\Omega}$, and the sample picture $\overline{M}$.

The background noise is subtracted from the reference and the sample picture. In this way the influences on the signal deriving from the technical equipment are eliminated. The final remission, where the inhomogeneous lighting is compensated, is calculated with equation 1:

$$\text{remission}: \overline{R} = \frac{\overline{M} - \overline{\Delta}}{\overline{\Omega} - \overline{\Delta}} \tag{1}$$

In detail:

$$\overline{R} = \begin{bmatrix} \dfrac{\mu_{11} - \delta_{11}}{\omega_{11} - \delta_{11}} & \cdots & \dfrac{\mu_{1n} - \delta_{1n}}{\omega_{1n} - \delta_{1n}} \\ \vdots & & \vdots \\ \dfrac{\mu_{m1} - \delta_{m1}}{\omega_{m1} - \delta_{m1}} & \cdots & \dfrac{\mu_{mn} - \delta_{mn}}{\omega_{mn} - \delta_{mn}} \end{bmatrix} \tag{2}$$

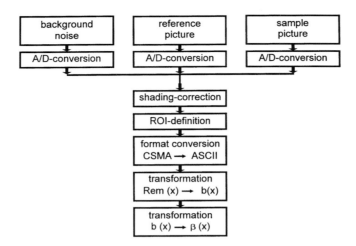

Figure 3 Shading correction for the compensation of inhomogeneous lighting.

By applying this shading correction (Equation 1 or 2) it is possible even to compensate changes in the positioning of the light sources. This was proven with experiments with an extremely inhomogeneous lighting, where this inhomogeneity was compensated by the shading correction.

After carrying out the shading correction the region of interest (ROI) has to be defined. This area is the part of the visualized sample which shall be quantified. Then the file format is transformed for the direct use of the data for graphics and the determination of the local surface mass density from the measured remission data by a calibration equation determined in experiments.

Calibration of the Visualization System

The remission data measured by optical measurements have to be transformed with a calibration equation to local mass transfer coefficient.

The calibration equation for the determination of the local surface mass density from the measured remission data is won experimentally. In these calibration experiments the foil or the membrane is colored homogeneously in small, stirred, cylindrical vessels. The foil is fixed at the inner side of the vessel and a water–dye solution in a certain concentration is filled in. The contents are stirred with a blade mixer. When the dye is transferred the stirrer is stopped and the colored foil is removed. The local surface mass density b can be calculated from the concentration of the dye solution and the area of the foil. This experiment is repeated to get a wide range of foils with different color intensities. The remission is measured either with the remission photometer or with the digital image processing system resulting in different calibration curves. The experimental data are approximated by a fifth-order polynomial which fits the data very well.

The local mass transfer coefficient β is calculated from the local surface mass density b by the following equation:

$$\beta = \frac{b}{c \cdot t} \qquad (3)$$

with c being the average concentration of the dye in the experimental apparatus. The decrease of the concentration while the experiment is running can be neglected, because the transferred amount of dye compared to the amount in the system is negligible.

With the local mass transfer coefficient β the local Sherwood number Sh can be determined, if the diffusion coefficient D is known (Equation 4).

$$Sh = \frac{\beta \cdot x}{D} \qquad (4)$$

In Equation 4, x represents a geometrical quantity; usually the hydraulic diameter is used. The diffusion coefficient D of Acidol–Blau is unknown. Therefore the diffusion coefficient was determined experimentally in a Stokes diaphragm cell (Kühnel and Kottke, 1994).

Applications

The technique for the visualization and determination of heat and mass transfer by chemisorption was applied in investigations of the cleaning behavior of multiple jet cleaning systems, flow phenomena in sinusoidal wavy channels, in spacer-filled flat channels, and plate heat exchangers.

Jet Cleaning Systems

In jet cleaning systems multiple jets are used for reducing the amount of CIP liquids in food processing plants or replacing the different kinds of chlorinated hydrocarbons by ecologically harmless detergent for the cleaning of machine parts in metal-producing industries. The cleaning behavior of the additives is combined with the power of impinging jets.

The removal by the impinging jets depends on the momentum transfer. By using the visualization technique presented here it is possible to visualize the mass transfer by the impinging jets. In experiments where the physical removal of a test layer was compared with the visualized mass transfer, it was shown that the color intensity distribution resulting from the visualization technique correlates to the momentum transfer at the removed test layer. Figure 4 shows the visualized mass transfer of three impinging jets.

Taylor-Goertler Vortices In Sinusoidal Wavy Channels

Wavy channels enhance the heat and mass transfer in many apparatuses. A flow instability observed in the concave part of the ducts leads to longitudinal vortices which are comparable to the Goertler vortices at concave walls or the Dean vortices in tube flow. The concave–convex walls in a sinusoidal wavy channel alternately enhance and destroy the longitudinal vortices resulting in highly complex flow and transport phenomena (Figure 5).

Spacer-Filled Channels

There are different applications for the use of spacers: supporting nets in membrane modules and turbulence promotors for the enhancement of heat and mass transfer. Additionally, spacers enhance the mixing and therefore they are used as static mixers.

Investigations were made by varying the shape of the cross section of the net wires, the inclination angle of the net wires, and the mesh size of the nets (Zimmerer and Kottke, 1994). The mixing behavior depends strongly on the inclination angle and is visualized by injecting an Acidol–Blau solution at a certain point at the beginning of the test section. In Figure 6 the chromatographic foil fixed at the wall shows that an excellent mixing behavior can be achieved.

Figure 4 Visualized mass transfer distribution of three impinging jets, using full cone nozzles. $H/D = 62.5$; $Re = 400{,}000$.

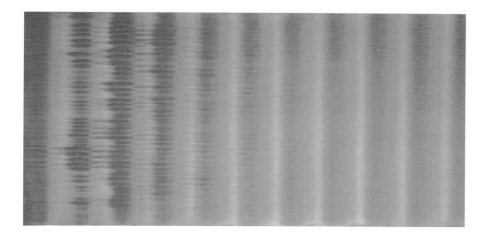

Figure 5 Taylor–Goertler vortices in a sinusoidal wavy duct. $Re = 557$; $s/a = 2.0$; $\lambda/a = 14.25$.

Plate Heat Exchangers

Crosswise corrugated plates are used for the enhancement of heat transfer in plate heat exchangers. Investigations (Gaiser and Kottke, 1994) concerning the heat transfer and the pressure loss are carried out by varying the amplitude a and the inclination angle λ. Figure 7 shows an example of the mass transfer distribution in a crosswise corrugated channel.

Conclusion

A method for the visualization and determination of heat and mass transfer at solid walls in liquid flow is presented. The method is based on chemisorption where the mass transfer is visualized as a locally and permanently fixed color intensity distribution. Even in complex flow fields, good results with a high local resolution are achieved. In different

Figure 6 Mixing behavior in a spacer-filled duct at Reynolds number $Re_h = 1000$ and an inclination angle $\alpha = 30°$. A dye tracer is injected at a certain point on the left side.

Figure 7 Mass transfer distribution in crosswise corrugated plates at $Re = 2000$. (Amplitude a = 1.825 mm, dimensionless wavelength $\lambda/a = 14.25$.)

experiments all equations necessary for the evaluation of mass transfer coefficients and Sherwood numbers are determined.

By using polyamide membranes for visualization, the flow field in apparatus with a trans-membrane flow can be visualized.

Acknowledgments

The financial support by the German DFG — Deutsche Forschungsgemeinschaft — and the support by the BASF AG (different dyes) are gratefully acknowledged.

Nomenclature

a	m	amplitude of a sinusoidal wavy channel
b	g/m²	surface mass density
c	g/m³	concentration
d	m	diameter
D	m²/s	diffusion coefficient
h	m	distance between nozzle and surface
R	—	remission
Re	—	Reynolds number
s	m	duct height

Sh	—	Sherwood number
t	s	time
x	m	geometrical parameter

Greek Symbols

β	m/s	mass transfer coefficient
δ	—	gray scale of the background noise
Δ	—	background noise matrix
λ	m	wavelength
μ	—	gray scale of the sample picture
M	—	sample picture matrix
ω	—	gray scale of the reference picture
Ω	—	reference picture matrix

Subscripts

| m | — | matrix counter: rows |
| n | — | matrix counter: columns |

References

Carlomagno, G. M., Flow Visualization and Heat Transfer Measurements by Infrared Thermography, in *Proc. 6th Int. Symp. Flow Visualization,* Yokohama, Tanida, Y., Miyashiro, H., Eds., Springer-Verlag, Berlin, 1992, 514.

Gaiser, G., Kottke, V., Plate Heat Exchangers for Food and Drink Processing — Aspects for Structure Optimization, in *Proc. Food Process Engineering,* Bath, 1994, 125.

Kottke, V., Blenke, H., Schmidt, K. G., Eine remissionsphotometrische Meßmethode zur Bestimmung örtlicher Stoffübergangskoeffizienten bei Zwangskonvektion in Luft, *Wärme Stoffübergang,* 10, 9, 1977.

Kottke, V., Kühnel, W., Becker, S., A New Technique for Visualization and Determination of Local Mass Transfer at Solid Walls in Liquid Flow, in *Proc. 6th Int. Symp. on Flow Visualization,* Yokohama, Tanida, Y., Miyashiro, H., Eds., Springer-Verlag, Berlin, 1992, 535.

Kühnel, W., Kottke, V., Visualization and Determination of Mass Transfer at Solid Walls in Liquid Flow, in *Proc. of FLUCOME 94,* Toulouse, 1994, 989.

Martin, H., Heat and Mass Transfer between Impinging Gas Jets and Solid Surfaces, *Adv. Heat Transfer,* 13, 1, 1977.

Zimmerer, C. C., Kottke, V., Flow Visualization and Mass Transfer in Spacer-Filled Flat Channels, in *Proc. FLUCOME 94,* Toulouse, 1994, 793.

chapter ten

Heat Transfer to Air from a Yawed Circular Cylinder

Gennaro Cardone[1], Guido Buresti[2], Giovanni Maria Carlomagno[1]

[1]DETEC, Università di Napoli, Naples, Italy
[2]DIA, Università di Pisa, Pisa, Italy

Abstract—Heat transfer distributions on a circular cylinder placed in a longitudinal flow at angles of attack α up to 20° are measured in a wind tunnel, for Reynolds number Re ranging from 84,700 to 159,000. The model is fitted with two different blunt forebodies, a flat sharp-edged disk or a hemispherical nose, and is constructed so as to perform temperature measurements by means of infrared thermography. The results show that for $\alpha = 0°$ the distribution of the Nusselt number on the surface of the sharp-edged cylinder is consistent with previous data obtained at lower values of Re. On the same cylinder, the leading edge separation bubble that is present for axisymmetric flow still persists with increasing α, but changes considerably in shape, due to a moderate increase of its length on the leeward side and a substantial decrease on the windward side. Conversely, all the data relative to the cylinder with the hemispherical nose do not show evidence of a separation bubble, probably because the latter is much smaller in size and located upstream of the measured region. As the angle of attack increases, longitudinal vortices begin to detach from the lateral regions of the model, but the different extent of the front separation bubble in the two cylinders leads to significant differences of the detachment features at corresponding angles of attack. In particular, in the case of the hemispherical-nosed cylinder, the vortices seem to appear at higher values of α, but then quickly move more upstream than in the sharp-edged case. Furthermore, the development of a quasi-two-dimensional flow field around the cylinder is always attained more upstream in the case of the cylinder with a hemispherical forebody. Finally, within the range of the present tests and in spite of the three-dimensionality of the flow, the variation of the Reynolds number does not seem to significantly affect the surface distributions of the parameter $Nu/Re^{0.75}$ for both model configurations.

List of Symbols

D   cylinder diameter
h   convective heat transfer coefficient
k   thermal conductivity of copper tracks
q_c tangential (to wall) conductive heat flux
q_j heat flux due to Joule heating
q_r radiative heat flux
Nu   Nusselt number (hD/λ)

Re	Reynolds number (VD/ν)
s	thickness of copper tracks
T_w	wall temperature measured by the radiometer
T_∞	free stream temperature
V	free stream velocity of air
x,y,z	cartesian coordinates
α	angle of attack
β	azimuthal coordinate
λ	thermal conductivity of air
ν	kinematic viscosity of air

Introduction

The study of the velocity field and of the surface heat transfer of bodies exhibiting regions of separated flow is of great interest from both the scientific and practical points of view. In particular, cylindrical bodies with circular cross section placed in a longitudinal flow are found in many engineering applications; the flow field around them is characterized by different types and extent of flow separation and reattachment according to the geometry of the cylinder upstream end and to the angle of attack of their axis relative to the incoming flow.

The flow field and surface heat transfer characteristics of a sharp-edged circular cylinder immersed in a stream at zero angle of attack α are the subject of a number of previous investigations. Ota (1975), from experiments carried out for the Reynolds number Re (based on the cylinder diameter) ranging from 24,900 to 53,600, finds that the reattachment point of the bubble originating at the front sharp leading edge occurs at 1.6 diameters from the latter (x/D = 1.6); he derives this value as the mean of measurements obtained by exploiting three different techniques (tuft exploration, zero skin friction, and nearly maximum pressure). However, in a subsequent paper concerning heat transfer measurements (Ota and Kon, 1977), it is argued that the flow reattachment occurs at the position where a maximum is attained for the heat transfer coefficient, i.e., at x/D = 1.3. Actually, the authors report the value 1.4, but Sparrow et al. (1987) notice that the effective maximum occurs at x/D = 1.3. For similar Reynolds numbers, the same position for the heat transfer maximum is found by Sparrow et al. (1987), who demonstrate, however, through surface visualizations that the real flow reattachment occurs further downstream, and in particular at x/D = 1.6 for the same flow conditions; in addition, they find that the values of both the locations at which flow reattachment and maximum heat transfer occur decrease for Reynolds number decreasing below the value of 30,000. It should be mentioned that the free-stream turbulence intensity is 0.4 to 0.5% in the tests of Sparrow et al., while it is approximately 0.8% in the tests of Ota.

More recently, detailed mean and fluctuating flow field measurements with hot wire anemometry in air are presented by Kiya et al. (1991) on a sharp-edged circular cylinder for Re = 200,000 and $\alpha = 0°$, with a free-stream turbulence intensity level of about 0.2%. The reattachment point is accurately derived from hot wire velocity measurements in the vicinity of the wall, as the location where the reverse-flow time fraction is equal to 0.5, and is also found to occur at x/D = 1.6. In another investigation in a water channel, Kiya et al. (1993) detect with LDV the reattachment point at x/D = 1.3 for Re between 4000 and 9200; this lower value is attributed to the higher turbulence level, which is 1.2 to 1.6%. However, this result seems even larger than the ones presented by Sparrow et al. (1987) with a lower turbulence intensity at similar Reynolds numbers, and this discrepancy remains yet to be explained. Finally, it should be mentioned that recent measurements (Carlomagno, 1995) of the heat transfer coefficient in a low-turbulence (0.08 to 0.12%) stream, for Re ranging from 16,300 to 32,100, show a maximum heat transfer occurring at x/D from 1.8 to 2.0, respectively, i.e., at much higher values than the ones presented by previous researchers. Again, the turbulence level seems to be the most probable cause of these differences. Indeed, the decrease of the length of the separation bubble with increasing turbulence intensity is well documented in the literature, even if the available data also suggest a reduction of this effect with increasing scale of the turbulence structures (Nakamura and Ozono, 1987; Simpson, 1989).

To the authors' knowledge, less information is available for cylinders at an angle of attack. In the case of a circular cylinder with a hemispherical nose, the evolution of the surface skin-friction lines with increasing incidence, at a Mach number equal to 1.2, is described by Peake and Tobak (1982). In the case of axisymmetric flow $\alpha = 0$, a separation bubble still exists immediately downstream of the junction between the hemispherical nose and the cylindrical body; however, this bubble is much smaller than the one in the sharp-edged case and the reattachment point seems to be located at $x/D \cong 0.15$. As the angle of attack increases, the bubble first becomes nonsymmetrical, and then tends to disappear from the windward side of the cylindrical part, so that for $\alpha > 5°$ a complex configuration of separation and reattachment lines develops, with the appearance in the flow of longitudinal-concentrated vortices, which persists up to very high values of α. Some more data on the topology of the three-dimensional flow around cylinders with a rounded nose are given by Tobak and Peake (1982).

Conversely, no systematic study of the evolution of the flow field or of the surface heat transfer seems to be available for sharp-edged cylinders at an angle of attack. Nevertheless, it is reasonable to expect that the separation imposed by the front sharp edge may lead to an extended range of α for which a separation bubble is still present in the vicinity of the leading edge, and that the subsequent development of the flow is also, in this situation, more complex than in the case of zero angle of attack.

In the present paper heat transfer measurements on a circular cylinder at angles of attack varying from 0 to 20° and for Reynolds numbers that extend the range of available data are performed by means of infrared thermography. The cylinder has either a flat sharp-edged or a hemispherical forebody attached to it. Infrared thermography is a nonintrusive technique and allows a rapid measurement of the surface heat transfer for various flow conditions. It is important to point out that, in the following, since the present work is concerned with heat transfer measurements, the term *separation bubble* is used with regard to the thermal distribution on the cylinder surface.

Experimental Setup

The experiments are carried out in the subsonic wind tunnel of the Department of Aerospace Engineering of Pisa University, which is a closed-return Göttingen-type wind tunnel with an open circular test section, 1.1 m in diameter and 1.5 m in length, and with free-stream turbulence of approximately 0.9%.

The experiments are performed within the range $84{,}700 < Re < 159{,}000$, with a test cylinder having a diameter of 80 mm and an overall length of 1100 mm. The model is supported at its back end by means of a sting, fastened to a special rig that allows the angle of attack to be varied continuously from –25 to +25°. Furthermore, in order to avoid any appreciable vibration of the cylinder, four thin steel wires connect a section of the model, aft of the measuring region, to a rigid frame positioned out of the wind-tunnel test section. Two different blunt forebodies may be attached to the cylinder: a flat, sharp-edged disk or a hemispherical nose having a radius of curvature equal to 40 mm.

For the employed testing conditions, since aerodynamic heating is not sufficient for the sensitivity of the infrared system, it is necessary to artificially create a temperature difference between the cylinder and the flow. For this reason, the model surface is made out of a printed circuit board so as to produce, by Joule effect, a *constant* heat flux over it. The

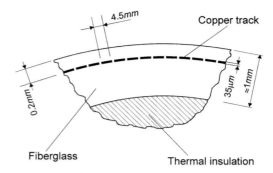

Figure 1 Sketch of the model cross section.

thickness and width of the copper-conducting tracks of the circuit, which are aligned with the cylinder axis, are manufactured with very close tolerances; in detail, the tracks are 35 µm thick, 4.5 mm wide, and placed at 5.0 mm pitch. The heated zone of the cylinder (i.e., the length of the copper tracks) is 600 mm long and starts 10 mm downstream of the sharp leading edge. The board (which constitutes the cylinder wall exposed to the free stream) has an overall thickness of about 1 mm, with the conducting tracks plunged 0.2 mm below its external surface, and is in contact with a thick layer of thermal insulation which is placed inside it. The sketch of the model cross section is shown in Figure 1. A stabilized DC power source supplies the electric current to the circuit and the power input is monitored by precisely measuring voltage drop and current across it.

The model surface, which is viewed by the infrared camera, is coated with a thin layer of black paint that has a directional emissivity coefficient equal to 0.95 in the wavelengths of interest and for viewing angles comprised between 40 and 90°; the data corresponding to viewing angles lower than 40° are not herein considered, since the directional emissivity coefficient of the surface decreases rapidly below this value.

The infrared camera measures the distribution of the wall temperature T_w of portions of the heated surface. The wall temperature is correlated to the heat transfer coefficient by means of the steady-state *heated-thin-foil* technique (Carlomagno and de Luca, 1989). In particular, for each pixel of the digitized thermal image, the local convective heat transfer coefficient h is calculated as:

$$h = (q_j - q_r - q_c)/(T_w - T_\infty) \tag{1}$$

where q_j is the Joule heat flux, q_r the radiative flux to ambient, q_c the tangential conduction heat flux along the cylinder wall, and T_∞ is the free stream temperature. Because of the low value of the pertinent Biot number, the heated wall may be considered as isothermal across its thickness. The radiative thermal losses q_r are computed by knowing the measured T_w, the total emissivity of the wall, and the ambient temperature. Under the assumption that the heat flux within the wall occurs mainly along the copper tracks (their thermal conductance is more than ten times larger than that of the fiberglass board), thermal losses due to tangential conduction are calculated by means of the Fourier law

$$q_c = -k\, s\, \frac{\partial^2 T_w}{\partial x^2} \tag{2}$$

where k is the copper thermal conductivity and s the thickness of the tracks. Radiative thermal losses turn out to be of the order of a few percent.

The heat transfer coefficients are evaluated in nondimensional form by means of the local Nusselt number:

$$Nu = h\,D/\lambda \tag{3}$$

where D is the diameter of the cylinder and λ the thermal conductivity coefficient of air evaluated at film temperature.

The employed infrared thermographic system is the AGEMA Thermovision 880. The field of view (which depends on the optical focal length and on the viewing distance) is scanned by the Hg–Cd–Te detector in the 8 to 12 µm infrared window. The nominal sensitivity, expressed in terms of noise equivalent temperature difference, is 0.07°C when the scanned object is at ambient temperature. The scanner spatial resolution is 175 instantaneous fields of view per line at 50% slit response function. A 20° lens is used during the tests at a viewing distance of 2 m, which gives a field of view of about 0.3×0.3 m²; this is done to obtain a higher spatial resolution in the thermal image. Therefore, in order to measure temperatures in the whole heated zone and to account for the directional emissivity coefficient, six different thermal images (two in the longitudinal direction times three in the azimuthal one) are taken and *patched up*. Each image is digitized at 8 bits in a frame of 140×140 pixels. An application software can perform: patching of images, identification of the measured points on the cylinder surface, noise reduction by numerical filtering, computation of temperature, and heat transfer correlations.

An error analysis, based on the calibration accuracy of the infrared system and repeatability of measurements, indicates that heat transfer measurements are accurate to within about ±5%.

Results

The fluid flow configuration and the coordinate axes used for the description of the results are schematically shown in Figure 2. It should be noted that the position $x = 0$ corresponds to the start of the cylindrical part of the body, so that it coincides with the leading edge of the flat, sharp-edged model, and with the end of the forebody in the case of the model with a hemispherical nose (which thus originates at $x/D = -0.5$). The azimuthal coordinate β starts at the leeward generating line. Furthermore, it has to be pointed out that, due to the effective position of the heating and to the strong end-conduction effects near the forebody and the afterbody, the portion of the cylinder for which the infrared camera gives reducible data actually starts at $x/D = 0.25$ and ends at $x/D = 7.5$.

A comparison between the results obtained in the present investigation for the sharp-edged cylinder (at zero angle of attack and for the three different Reynolds number values) and those obtained by Ota and Kon (1977) is shown in Figure 3. The whole data are reported in terms of the parameter $Nu/Re^{0.75}$ already used by Ota and Kon which, as can be seen from the figure, allows a good coalescence of all the present measurements. The axial distribution of this parameter is characterized by an initial increase, a local maximum, which is located at approximately $x/D = 1.2$, and a subsequent monotonic decrease corresponding to a redeveloping flow. The initial low heat transfer region corresponds to the separated flow which is present just downstream of the cylinder nose. It must be remembered that the location of the maximum convective heat transfer does not exactly coincide with that of flow reattachment, which is located slightly downstream of this position (Sparrow et al., 1987).

As can be seen from Figure 3, the position of the maxima and their values are in all cases relatively consistent with the findings of Ota and Kon (1977) that are also shown in the figure, as well as with the data (not shown in Figure 3) of Sparrow et al. (1987) for Re

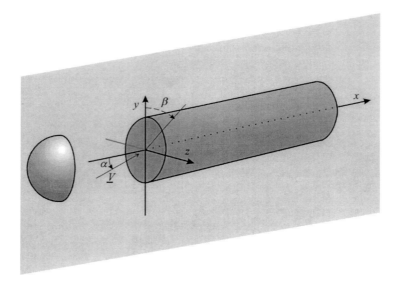

Figure 2 Fluid flow configuration and coordinate axes.

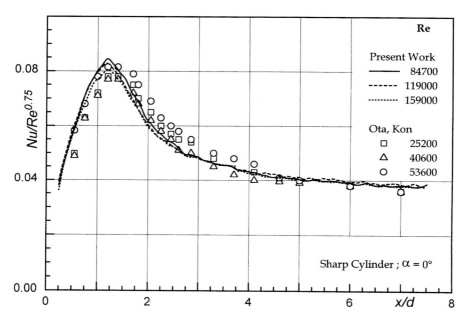

Figure 3 Distribution of the dimensionless heat transfer coefficient on the sharp-edged cylinder at α = 0 for various Reynolds numbers.

> 30,000, notwithstanding the different ranges of Re. The present data show a sharper peak with respect to that of Ota and Kon; this difference may be most certainly ascribed to the fact that Ota and Kon seem to ignore tangential conduction effects in their relatively thick-walled model.

The corresponding data for the cylinder with hemispherical nose at zero angle of attack are shown in Figure 4. Also, in this case the dimensionless group $Nu/Re^{0.75}$ makes all the curves to practically coincide into a single one; however, no heat transfer maximum is

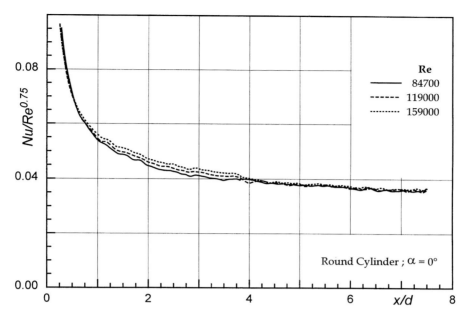

Figure 4 Distribution of the dimensionless heat transfer coefficient on the cylinder with hemispherical nose at $\alpha = 0$ for various Reynolds numbers.

found in this case, the curves showing a continuous monotonic decrease with the increasing of the longitudinal coordinate. From this it can be inferred that the measuring zone, which excludes $x/D < 0.25$, is most likely positioned downstream of the reattachment of the smaller separation bubble that has been found immediately behind the end of the hemispherical portion by Peake and Tobak (1982).

By comparing the data of Figures 3 and 4, it may be noticed that, at high values of x/D, the heat transfer coefficients for the sharp edge case are slightly larger than the ones for the cylinder with hemispherical nose; this is probably due in part to the fact that the round-nosed cylinder starts $0.5\ D$ ahead of the sharp one and in part to the higher turbulence level induced by the latter.

In Figures 5 to 8 the evolution of the isocontours of the dimensionless surface heat transfer coefficient $Nu/Re^{0.75}$ is shown in the case of the sharp-edged model at four increasing angles of attack for Re = 119,000. To highlight the surface flow features, the dimensions of the cylinder are not shown to scale; in fact, the represented cylinder length is equivalent to $7.25\ D$ and, as already mentioned, it starts at $0.25\ D$ from the leading edge. The different values of the parameter $Nu/Re^{0.75}$ may be derived from the color scale attached to the figures, which also show the position of the windward generating line corresponding to $\beta = 180°$.

From Figures 5 to 8 it can be noticed that as α increases, the heat transfer surface pattern becomes progressively more two-dimensional. In particular, it is clear that the shape of the separation bubble, whose presence may still be recognized at the lower angles of attack, changes with a strong reduction of its length on the windward side and a slight increase on the leeward side. Furthermore, inside the bubble the intensity of the heat transfer varies azimuthally, with a significant decrease in the upper region (i.e., toward $\beta = 0°$) with respect to the case of $\alpha = 0°$. This feature is probably a consequence of the reduction of the friction coefficient connected with the coalescence of the limit streamlines toward the leeward generating line in the plane of symmetry that, by using the terminology of Tobak and Peake (1982), becomes a *local line of separation*. For $\alpha = 20°$, the highest

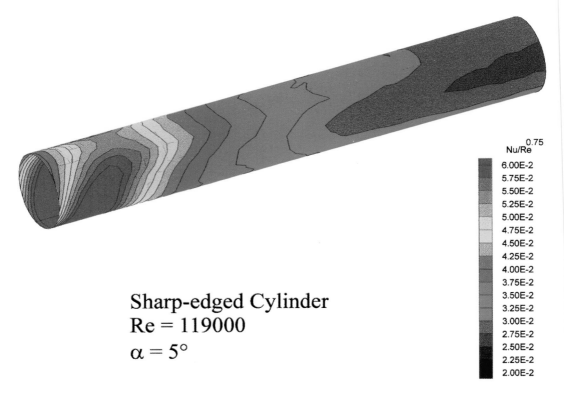

Figure 5 Isocontours of dimensionless heat transfer coefficient on the sharp-edged cylinder at $\alpha = 5°$, Re = 119,000.

heat transfer zones are reduced to two small spots located in the vicinity of $\beta \cong 90°$ at the beginning of the measured region, and, by extrapolating upstream the heat transfer map, the separation bubble at $\beta = 180°$ (windward generating line) seems to disappear.

Another feature that can be derived from Figures 5 to 8 is the progressive appearance, on the cylinder sides, of a low heat transfer region, which starts in the aft end of the image at $\alpha = 5°$, and moves upstream with increasing angle of attack, simultaneously becoming sharper and seemingly enhancing the global three-dimensionality of the flow. This region is presumably connected with the fact that the increasingly intense cross-flow leads first to instabilities of the boundary layer, and eventually to the separation from the sides of the cylinder of dominating longitudinal vortical structures, similar to those described by Tobak and Peake (1982) for a round forebody. Within the heated region, the azimuthal location of the detected minimum heat transfer coefficient varies from $\beta \cong 70°$ for $\alpha = 5°$, to $\beta \cong 55°$ for $\alpha = 10°$ and $15°$, to $\beta \cong 75°$ for $\alpha = 20°$.

The curves of Figure 9 give a clear picture of the remarkable quantitative effect of the three-dimensionality of the flow along the leeward generating line. Indeed, it is apparent that even small angles of attack cause a sharp reduction of the heat transfer inside the separation bubble at $\beta = 0°$. Conversely, further increases of α affect only slightly the trends and values of Nu, with a limited downstream migration of the location corresponding to the heat transfer maximum, whose value remains almost unchanged.

The map of the dimensionless surface heat transfer coefficient $Nu/Re^{0.75}$ for $\alpha = 10°$ and Re = 84,700 is shown in Figure 10. By comparing this figure with Figure 6 it is possible to notice that, in this case, the variation of Reynolds number seems to influence only marginally the isocontours of $Nu/Re^{0.75}$ on the surface of the flat-nosed cylinder even at an angle of attack. It should be noted that this is not an obvious result, as the fact that the considered

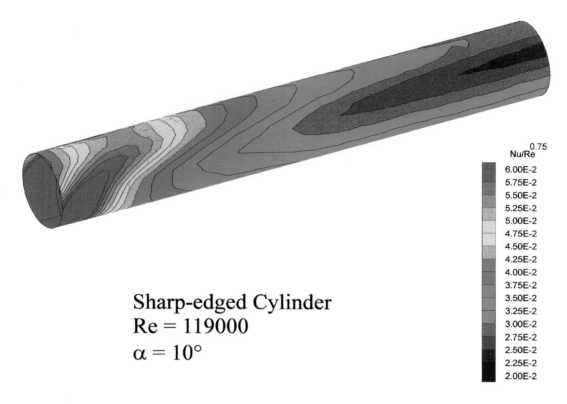

Figure 6 Isocontours of dimensionless heat transfer coefficient on the sharp-edged cylinder at α = 10°, Re = 119,000.

dimensionless parameter may take into account all the influence of the Reynolds number variation has previously been found only for the case of axisymmetrical flow. The extension of this behavior to the three-dimensional case is not straightforward. However, it should also be pointed out that the constancy of $Nu/Re^{0.75}$ with varying Reynolds number may be expected to hold only if the consequent changes in the boundary layer conditions do not lead to significant modifications of the topology of the surface flow, i.e., if there are no variations in the singular points of the skin-friction patterns and in the separation lines, as indeed seems to be the case of Figures 6 and 10.

Figures 11 to 13 show the evolution of the isocontours of the dimensionless surface heat transfer coefficient $Nu/Re^{0.75}$ in the case of the cylinder with a hemispherical forebody at three increasing angles of attack for Re = 119,000 (the map relative to α = 10° has not been shown since the pertinent data are not complete over the whole cylinder surface). By comparing these maps with the ones of Figures 5, 7, and 8 it is possible to notice that the change from the sharp-edged forebody to the hemispherical one causes significant differences to occur, from both the qualitative and the quantitative points of view. Indeed, the overall heat transfer is lower than for the sharp-edged cylinder at corresponding angles of attack and the separation bubble now seems to be either absent or extremely reduced in size, and in any case presumably positioned immediately before the zone viewed by the infrared camera.

The influence of the leading edge on the surface heat transfer, and consequently on the flow field, is much more confined to the initial region of the cylinder. Furthermore, there is a much more rapid progression of the surface pattern, and consequently of the flow features, with increasing angle of attack; in particular, the lateral region of low Nu value, linked to the separation of a longitudinal vortex which is absent for α = 5°, appears for

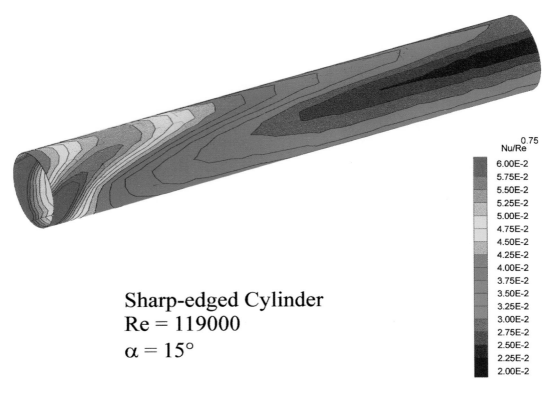

Figure 7 Isocontours of dimensionless heat transfer coefficient on the sharp-edged cylinder at $\alpha = 15°$, Re = 119,000.

$\alpha = 10°$ and rapidly moves upstream, dominating the whole side region for $\alpha = 20°$. Within the heated region, the azimuthal location of the detected minimum heat transfer coefficient varies from $\beta \cong 55°$ for $\alpha = 10°$, to $\beta \cong 80°$ for $\alpha = 15°$ and 20°. Therefore, by comparing Figures 5, 7, and 8 with Figures 11 to 13, the conclusion can be drawn that in the case of the sharp-edged cylinder the persistence and considerable extent of the front separation bubble inevitably produce a downstream movement of the longitudinal vortices separating from the sides of the cylinder at high angles of attack, whereas with the smoother hemispherical forebody the detachment of these vortices occurs more upstream for the same values of α, moving rapidly to a position just behind the start of the cylindrical region of the model. This behavior is also clearly shown by the almost constant (in the x direction) heat transfer coefficient on the leeward side of the round-nosed cylinder aft part. All the above-described findings show a more upstream located attainment of a quasi-two-dimensional flow field, typical of cylinders in cross-flow, in the case of the cylinder with a hemispherical nose.

Finally, also in the case of the hemispherical-nose model and within the analyzed data, the Reynolds number seems to influence only slightly the main flow features, which vary essentially in the initial part of the cylinder. This behavior is shown in the curves of Figure 14, which demonstrate that the trends (for varying Reynolds number) of the heat transfer parameter $Nu/Re^{0.75}$ are remarkably similar along the leeward generating line ($\beta = 0°$) of the fore part of the cylinder at $\alpha = 10°$, with only a presumable moderate upstream movement of the maximum value, which, however, is not covered by the measurements.

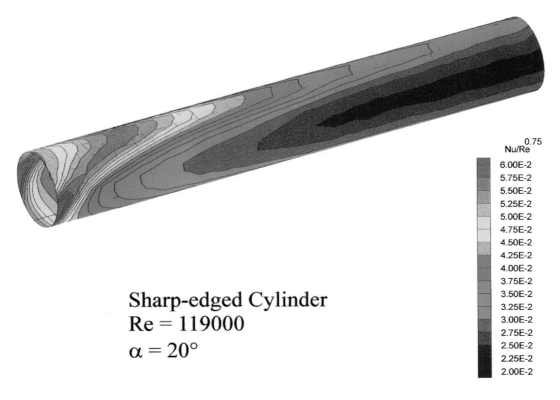

Figure 8 Isocontours of dimensionless heat transfer coefficient on the sharp-edged cylinder at $\alpha = 20°$, Re = 119,000.

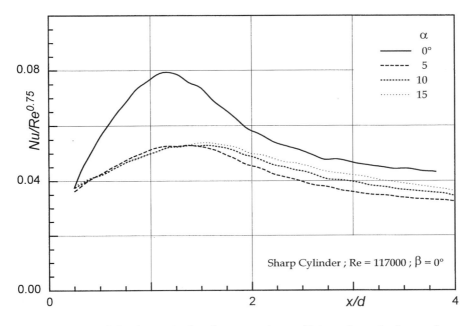

Figure 9 Distribution of the dimensionless heat transfer coefficient along the leeward generating line ($\beta = 0$) on the sharp-edged cylinder at various angles of attack and Re = 119,000.

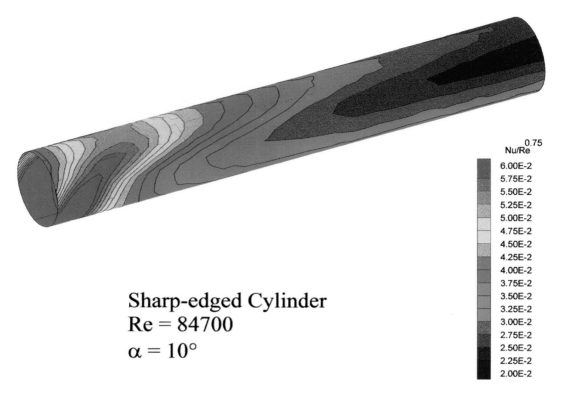

Figure 10 Isocontours of dimensionless heat transfer coefficient on the sharp-edged cylinder at $\alpha = 10°$, Re = 84,700.

Conclusions

The present paper shows that infrared thermography may be an adequate means to rapidly obtain significant information on the surface heat transfer and flow field characteristics of different cylindrical models positioned in a stream at increasing angles of attack, provided the models are suitably constructed in order to allow the heated-thin-foil technique to be implemented.

In particular, this technique is applied to study the two-dimensional heat transfer distribution on the surface of two cylinders, having either a flat, sharp-edged or a hemispherical forebody. The cylinder models are placed in a wind tunnel having a turbulence intensity of 0.9% at an angle of attack varying from 0 to 20° and for Reynolds number comprised between 84,700 and 159,000.

The results for the sharp-edged cylinder at $\alpha = 0°$ show a one-dimensional distribution of the Nusselt number on the cylinder surface that is generally consistent with previous data obtained by other authors for lower values of the Reynolds number. In particular, the presence of a leading edge *thermal* separation bubble may be recognized, its length being determined by the position of the maximum heat transfer coefficient. When the angle of attack is increased, this bubble still seems to be present, but changes considerably in shape with a moderate increase of its length on the leeward side and a substantial decrease on the windward side. The increased three-dimensionality of the flow pattern is further enhanced by the appearance of low heat transfer regions on the aft sides of the model, seemingly connected with the detachment of longitudinal vortices, which gradually move upstream with increasing α.

Chapter ten: Heat transfer to air from a yawed circular cylinder

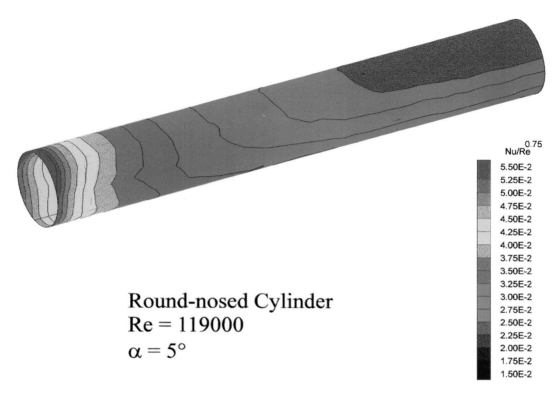

Figure 11 Isocontours of dimensionless heat transfer coefficient on the cylinder with hemispherical nose at α = 5°, Re = 119,000.

Conversely, the cylinder with the hemispherical nose is characterized by a decrease of the three-dimensionality of the flow in the body front region with respect to the previous case and probably by a much smaller separation bubble, positioned upstream of the measured region. Furthermore, the lateral longitudinal vortices seem to appear at a higher angle of attack, but then move quickly more upstream than in the flat-nose case, eventually completely dominating the flow pattern. Thus, the different extent of the separation bubble in the two cases, due to the different forebody geometries, leads to significant differences in the development of the flow with increasing angles of attack. The most significant difference is that a quasi-two-dimensional flow field around the cylinder is attained more upstream in the case of the cylinder with a hemispherical forebody.

Finally, for the analyzed data the variation of the Reynolds number does not seem to significantly affect the surface distributions of the parameter $Nu/Re^{0.75}$, which appears to be capable of taking this variation into account, in spite of the three-dimensionality of the flow. However, this result may be connected with the fact that within the analyzed Reynolds number range, the variations of the boundary layer on the two models do not significantly affect the general topology of the flow, and in particular the locations of the separation and reattachment lines.

Acknowledgments

The authors express their sincere thanks to Dr. Carosena Meola and Dr. Giovanni Lombardi for assistance in performing the experiments and to Dr. Luca Marino for assistance in reducing the data.

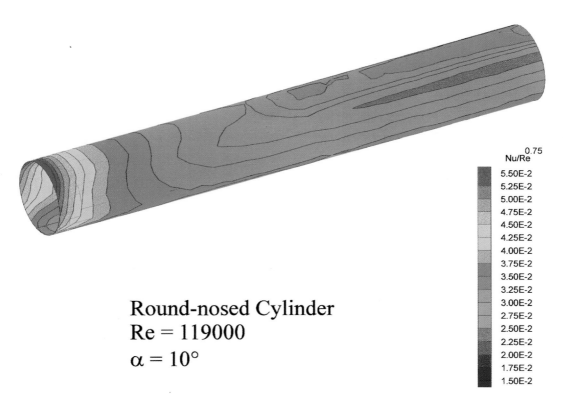

Figure 12 Isocontours of dimensionless heat transfer coefficient on the cylinder with hemispherical nose at $\alpha = 15°$, Re = 119,000.

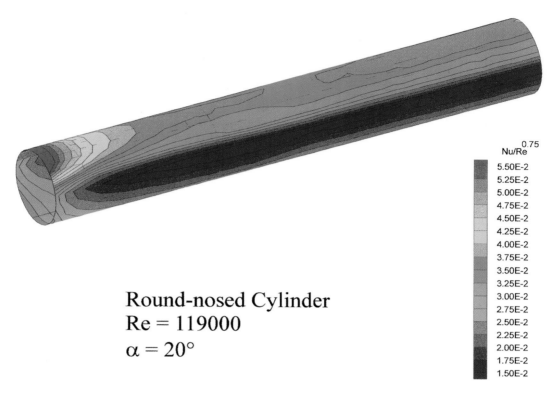

Figure 13 Isocontours of dimensionless heat transfer coefficient on the cylinder with hemispherical nose at α = 20°, Re = 119,000.

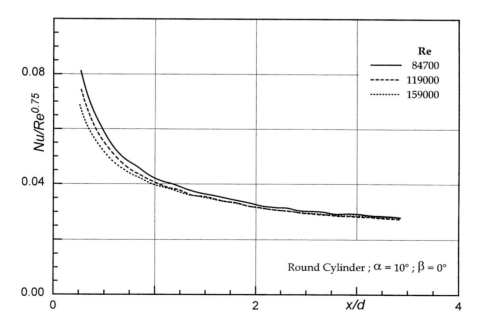

Figure 14 Distribution of the dimensionless heat transfer coefficient along the leeward generating line (β = 0) on the cylinder with hemispherical nose at α = 10° for various Reynolds numbers.

References

Carlomagno G.M. Heat Transfer Measurements and Flow Visualization Performed by Means of Infrared Thermography, Proc. Eurotherm Sem. No. 46, Heat Transfer in Single Phase Flows 4, Di Marco P. Ed., pp. 45–52, Pisa, 1995.

Carlomagno G.M., de Luca L. Infrared Thermography in Heat Transfer, in *Handbook of Flow Visualization*, Yang W.J. Ed., pp. 551–553, 1989.

Kiya M., Mochizuki O., Tamura H., Nozawa T., Ishikawa R., Kushioka K. Turbulence Properties of an Axisymmetric Separation-and-Reattaching Flow, *AIAA J.*, Vol. 29, pp. 936–941, 1991.

Kiya M., Mochizuki O., Ido Y., Kosaku H. Structure of Turbulent Leading-Edge Separation Bubble of a Blunt Circular Cylinder and Its Response to Sinusoidal Disturbances, *J. Wind Eng. Ind. Aerodyn.*, Vol. 49, pp. 227–236, 1993.

Nakamura Y., Ozono S. The Effects of Turbulence on a Separated and Reattaching Flow, *J. Fluid Mech.*, Vol. 178, pp. 477–490, 1987.

Ota T. An Axisymmetric Separated and Reattached Flow on a Longitudinal Blunt Circular Cylinder, *J. Appl. Mech.*, Vol. 42, pp. 311–315, 1975.

Ota T., Kon N. Heat Transfer in an Axisymmetric Separated and Reattached Flow Over a Longitudinal Blunt Circular Cylinder, *J. Heat Transfer*, Vol. 99, pp. 155–157, 1977.

Peake D.J., Tobak M. Three-Dimensional Flows About Simple Components at Angle of Attack, AGARD-LS-121, paper 2, 1982.

Simpson R.L. Turbulent Boundary-Layer Separation, *Annu. Rev. Fluid Mech.*, Vol. 21, pp. 205–234, 1989.

Sparrow E.M., Kang S.S., Chuck W. Relation Between the Points of Flow Reattachment and Maximum Heat Transfer for Regions of Flow Separation, *Int. J. Heat Mass Transfer*, Vol. 30, pp. 1237–1246, 1987.

Tobak M., Peake D.J. Topology of Three-Dimensional Separated Flows, *Annu. Rev. Fluid Mech.*, Vol. 14, pp. 61–85, 1982.

chapter eleven

A Visualization Study on Water Spray of Dragon Washbasin

Qing-Ding Wei, Da-Jun Wang, Bin Yan, Xiang-Dong Du, and Jun Chen

Department of Mechanics, SKLTR, Peking University, Beijing, China

> *Abstract*—"Dragon Washbasin Phenomena" refers to the water in a Dragon Washbasin — a circular basin that can spray water by rubbing it— which was studied by use of the particle image velocimetry technique and other visualization methods. Instead of using hands to rub the circular Dragon Washbasin, two vibrators were used. Along with change of the vibration frequency the surface wave and the velocity fields of the water in the basin were measured and visualized. The experimental results suggest that when the vibration frequency is far away from the natural frequency of the basin with water, the shape of the surface wave is regular and spray does not occur; when the vibration frequency is near the natural frequency the surface wave becomes irregular and spray occurs. Although the seiche wave can be excited in the meantime the spray will be checked.

Introduction

"Dragon Washbasin" is one of the marvels of Chinese ancient scientific discovery and is also a great art treasure. It is a mythical vessel made of bronze and in the form of a wash basin with two handles on its edge as shown in Figure 1. Some patterns of dragons or fishes are cast on the inside bottom of the basin. Therefore it is called "Dragon Washbasin" or "Fish Washbasin". When water is half filled in a Dragon Washbasin and the handles are rubbed appropriately with wet hands, a myriad of water spray emanates from some regular places near the basin wall, being just like dragons or fishes playing with water, as shown in Figure 2. The phenomena are very marvelous and dramatic and may be called "Dragon Washbasin Phenomena". Some studies have been done to clarify its mechanism. Wang (1991, 1993) has studied this both theoretically and experimentally. He found that Dragon Washbasin Phenomena are caused by self-excited vibration of the coupled system of Dragon Washbasin and water in it. The self-excited vibration may occur near the natural frequencies of the system. The regions where water emanates significantly are near the wave crests of the vibration mode shape of the vessel. For example, the mode shape of the first-order vibration of the side wall of a Dragon Washbasin is approximately $\sin 2\theta$; θ is the circular angle and is calculated from the symmetric line of the two handles of the basin. The positions of wave crests are at $\theta = (2k + 1)\pi/4$, $k = 0, 1, 2, 3$. Corresponding to the mode shape of the third-order vibration, being approximately $\cos 3\theta$, θ is equal to $k\pi/3$, $k = 0, 1, 2, \ldots 5$. Wang also found that Dragon Washbasin Phenomena can reappear by horizontal

Figure 1 The "Dragon Washbasin" — a great Chinese ancient art treasure.

forced vibration. Using commercially available upright circular cooking utensils and vibrators with adjustable frequency, he reproduced essentially and observed many interesting phenomena, which are difficult to find in the Dragon Washbasin Phenomena caused by hands rubbing. For example, the shape of the surface wave changes along with change of the vibration frequency; when the vibration frequency slightly exceeds the natural frequency of the system, seiche wave can be excited while water spray is checked; and so on. Besides, some computational studies concerned with Dragon Washbasin Phenomena have been conducted. A linearized theory of capillary–gravity waves in a circular basin has been developed by Shen et al. (1993). Some studies on nonlinear theory have been done by Hsieh (1994). It seems that there is a long road to go because theoretical studies face a lot of difficulties on the nonlinear problems concerned with Dragon Washbasin Phenomena.

This paper will introduce an experimental study on Dragon Washbasin Phenomena using the particle image velocimetry technique and other visualization methods to clarify the flow structure accompanying Dragon Washbasin Phenomena.

Figure 2 The "Dragon Washbasin Phenomena" — water spray caused by hands rubbing.

Experimental Apparatus and Method

Experimental Model

The experimental model was a circular glass basin with an upright side wall and with diameter 250 mm, height 180 mm, and wall thickness 5 mm, as shown in Figures 3 and 4. Each order natural frequency is as follows:

first order	second order	third order
312 Hz	708 Hz	1.22 kHz

When an experiment was conducted the glass basin would be filled with water, depth 100 mm. In such cases the system consisting of the glass basin and the water had the following natural frequencies:

first order	second order	third order
268 Hz	628 Hz	1.1 kHz

Figure 3 The experimental setup.

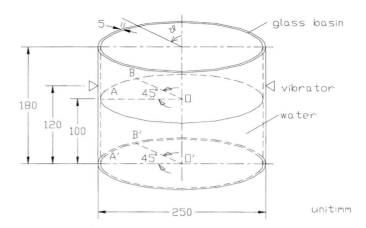

Figure 4 The testing basin.

Vibration Device

The vibration device used in this experiment consists of vibrators, signal generator, power source, and power amplifier. Two minivibrators with maximum amplitude of 3 mm and maximum power input of 15 VA were installed symmetrically at two sides of the basin and were placed at the height 120 mm from the bottom (see Figure 3).

Light Source

An argon-ion laser source with power 5 W and an optical system including a cylinder lens was used for providing a laser light sheet. A pulse regular system could adjust the laser light sheet as a pulse light sheet. Also, an ordinal incandescent light source with spotlight was also used.

Method of Fluorescence Induced by Laser

The fluorescer, fluorescein sodium, was used to visualize the flow patterns of Dragon Washbasin Phenomena. Whenever the laser sheet shines the water with fluorescein sodium green light will be induced.

Particle Image Velocimetry (PIV)

Tracing Particles: The nylon sphere-like particles with diameter 75 to 150 µm were used for tracing the flow on and under the surface of the water in the testing basis. The density of the particles is about $10^3/cm^3$.

Pulse Laser Light Sheet: Two pulse laser light sheets with interval 0.1 s and pulse width 0.005 s were used to expose film two times. The thickness of the light sheet is about 1 mm.

Particles Image Processing: To process the film recording the information of the tracer particles the following steps were adopted:

1. Using a video camera, scan the pictures developed and printed with the films and input the gray scale function $I_1(x,y)$ to a computer, x and y coordinates.

2. Pretreating $I_1(x,y)$ with the procedures of numerically filtering, eliminate noise and sharp information, and get new gray scale function $I_2(x,y)$.
3. Using Fourier transformation and inverse Fourier transformation, get autocorrelation function $J(\xi,\zeta)$.
4. On the basis of the peak values of function $J(\xi,\zeta)$, determine velocity vectors.

Results

The Relationship between Flow Patterns and the Vibration Frequency

The Vibration Frequency Much Lower Than the First-Order Natural Frequency of the Basis-Water System (268 Hz): In such cases, for example, the vibration frequency was lower than 200 Hz; the water in the basin was very quiet. There were not only obvious surface waves, but also no inner flow.

The Vibration Frequency Close to the First-Order Natural Frequency: When the vibration frequency was increased over about 230 Hz, an annular region next to the side wall had a lot of thin ellipse-like capillary waves whose long axes were basically perpendicular to the side wall. They were stable initially and could be seen as standing waves. When the vibration frequency was very near the first-order natural frequency, these ellipse-like waves became unstable, deformed, and eventually broke up into scalelike ripple. The closer to an excited point, the stronger and more unstable the ripple was. Figures 5(a) to 5(c) show the surface wave under the condition of the vibration frequency 253, 259, and 263 Hz, respectively. When the vibration frequency exceeded 260 Hz, a few water drops were splashed from the region near excited points.

The Vibration Frequency Equal to the First-Order Natural Frequency: In the time of vibration frequency reaching 268 Hz, the first-order frequency, a lot of water drops were splashed and formed spray. This appearance was exactly like the Dragon Washbasin Phenomena. The sprays occurred mainly at four regions, two of which were near excited points; the other two were at the positions perpendicular to the excited points. Therefore, four identical fan-shaped regions were formed. Figure 6 shows the four regions whose boundaries were formed by the foam of the spray. Figures 7 to 9 offer the results measured with the PIV technique. Figure 7 shows the distribution of the velocity vectors on a part region of the water surface. It can be seen that the flow comes together from spray regions to the center of the basis. There are some vortices at the central region. The values of velocity near the boundaries between two fan-shaped regions are small. Figures 8 and 9 show the distributions of the velocity vectors at two vertical sections, respectively, one of which passes the central line of a fan-shaped region; another passes its boundary. It is clear that there was a vortex ring in the basin in which an annular velocity field was formed at every vertical cross section.

The Vibration Frequency Slightly Over the First-Order Natural Frequency: When vibration frequency was increased just a little over the first-order natural frequency, for example, reaching 272 Hz, one kind of seiche wave would occur. Its wave crests were always at both the center and the side wall of the basis. Figure 10 shows the shape of the seiche wave excited. It can be seen that in such cases spraying was checked. This phenomenon was not found in the Dragon Washbasin Phenomena by hands rubbing.

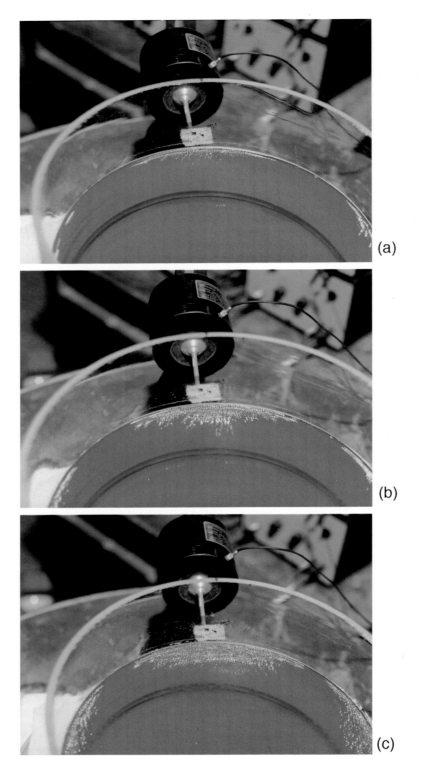

Figure 5 The surface wave under the conditions of various vibration frequencies. (a) 253 Hz; (b) 259 Hz; (c) 263 Hz.

Figure 6 The four-fan-shaped spraying region.

The Vibration Frequency Significantly Over the First-Order but near the Second-Order Natural Frequency: If the vibration frequency was far away from both the first- and second-order frequency, the water was very quiet. But if it was close to the latter, all of the phenomena occurring along with change of the vibration frequency would almost repeat that described above. Only the number of the spray regions became six as shown in Figure 11.

Conclusions

1. The Dragon Washbasin Phenomena occur when the vibration frequency is close to the natural frequency of the coupled system of the basin and the water filled in it.
2. The water spray is at four or six identical fan-shaped regions corresponding to the vibration frequency being near the first or second natural frequency of the basin-water system, respectively.
3. There is a vortex ring, the axis of which is the same with that of the circular basin when the Dragon Washbasin Phenomena occur.
4. The seiche wave can be excited if the vibration frequency exceeds slightly the natural frequency of the basin-water system.

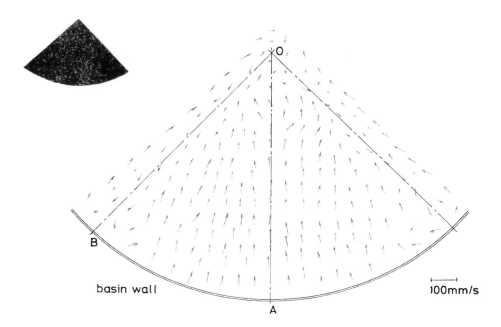

Figure 7 A photograph of the tracer particle image of the water surface (above) and the velocity vectors (below, about the positions of point O, A. B; refer to Figure 4).

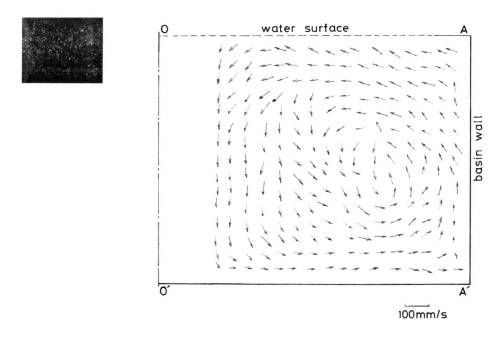

Figure 8 A photograph of the tracer particle image in a vertical section passing thorough the excited points (above) and the velocity vectors (below, about the positions of point O, O', A', A; refer to Figure 4).

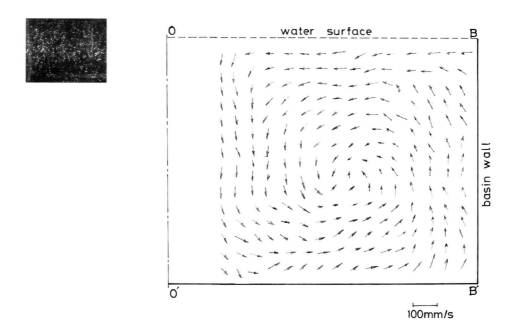

Figure 9 A photograph of the tracer particle image in a vertical section passing through the boundary of fan-shaped regions (above) and the velocity vectors (below, about the positions of point O, O', B', B; refer to Figure 4).

Figure 10 The seiche wave. (a) Wave crest; (b) equilibrium position; (c) trough.

Figure 11 The six fan-shaped spraying regions.

References

Hsieh, D. Y., Standing water waves in a circular basin, in *Proc. Int. Conf. Hydrodynamics*, China Ocean Press, Wuxi, China, 1994, 74.

Shen, M. C., Sun, S. M., and Hsieh, D. Y., *Wave Motion*, 18, 401, 1993.

Wang, D., A research on the mechanical properties of cultural relics of China, in *Proc. Scientific and Technological Archaeology*, Chinese University of Science and Technology Publishing House, Hefei, 1991, 51.

Wang, D., Study on mechanical characteristic of ancient cultural relics, *Sci. Conserv. Archaeol.*, 5(1), 35–39, 1993.

chapter twelve

Dynamics of Propane Jet Diffusion Flames

V. R. Katta[1], L. P. Goss[1], W. M. Roquemore[2], L.-D. Chen[3]

[1]Innovative Scientific Solutions, Inc., Dayton, Ohio
[2]Wright Laboratory, WL/POSC, Wright Patterson AFB, Ohio
[3]Department of Mechanical Engineering, The University of Iowa, Iowa City, Iowa

Abstract—*Jet diffusion flames differ from corresponding cold jets in several aspects; for example, the flames have longer potential cores and undergo slower transition to turbulence. Understanding the processes that make the flames behave differently from the cold jets helps develop accurate mathematical models for practical combustors. By using reactive-Mie-scattering laser techniques details of the flow structures of both the combusting and cold flows are obtained. These flow-visualization images suggest that the vortices inside the flame remain coherent for a long time and undergo structural changes as they convect downstream. Using a third-order accurate, time-dependent, axisymmetric numerical code these flows are simulated and the code is tested for its ability to predict the differences observed between the reacting and the cold jets. Numerical experiments are performed to understand the mechanisms responsible for these differences. Based on the numerical results an explanation is provided for the longer coherence lengths for vortices in flames. Experiments and numerical simulations are also made for flames at higher jet velocities to investigate the turbulent structure of a jet diffusion flame.*

Introduction

The understanding of jet diffusion flames is important in many practical applications and for developing theories of combusting processes. Because of this, they have been actively studied since the classic works published in the *3rd Symposium on Combustion Flame and Explosion Phenomena* in 1949 (Wohl et al., 1949; Hottel and Hawthorne, 1949; Hawthorne et al., 1949; Wohl et al., 1949). Considerable data on statistical quantities such as time-averaged and rms values of velocity, temperature, and species concentrations have been obtained with single-point measurement techniques (Bilger, 1976; Eickhoff, 1982; Faeth and Samuelson, 1986). These data have formed the bases for understanding many of the processes occurring in jet diffusion flames. Because of the success of the statistical approach there is a tendency to think about combustion processes in terms of time-averaged parameters. For engineering applications, there is definite value and, in many cases, a necessity of thinking in terms of mean values of parameters. However, there is a danger to this line of thinking in that the mean and fluctuating quantities can, in many cases, mask the physics and chemistry that are germane to understanding the fundamental processes that give rise to the statistical results. This is particularly true for near-laminar and transitional jet flames in which the impact of large-scale, organized, buoyancy-induced vortices on the air side of the flame and the Kelvin–Helmholtz-type vortex structures on the fuel side of the flame dominate the flame characteristics (Yule et al., 1981; Eickhoff and Winandy, 1985; Coats and Zhao, 1988; Roquemore et al., 1989; Davis et al., 1991). To gain an insight into these processes, it is helpful and perhaps essential to think in terms of the dynamic characteristics of jet flames.

This paper attempts to develop a qualitative view of the dynamic processes responsible for some of the important physics occurring in propane jet diffusion flames. The approach is to examine the structure of reacting and nonreacting jet flows experimentally and then to use a direct numerical simulation to probe the physics of the dynamic processes in these flows.

Experimental Methods

Experimental Setup

The experimental setup consists of vertically mounted coannular jets and is described in Chen and Roquemore (1986). Unless otherwise stated, the central fuel jet is a 25.4-mm-diameter tube which contracts to a 10-mm-diameter nozzle. The nozzle is designed to provide a flat mean velocity profile with low velocity fluctuations at the nozzle exit. The annulus air jet has a diameter of 245 mm. An air velocity of 0.15 m/s was found to be sufficient to reduce the room air disturbances in the first 15 diameters of the jet exit while not causing a significant effect on the visible flame structure.

Reactive-Mie-Scattering (RMS) Technique

The RMS technique provides a more detailed view of the mixing processes than is normally obtained by the shadowgraph and smoke (performed particle) visualization techniques (Chen and Roquemore, 1986). It involves seeding both the fuel and the dry annulus air with $TiCl_4$ vapor. The $TiCl_4$ reacts spontaneously and nearly instantaneously and nearly isothermally with the water product of the flame to form micron-size TiO_2 particles. The Mie scattering from these particles provides a view of where: (1) the water product is *molecularly* mixed with the air outside the flame surface, (2) the fuel is mixed with the water product inside the flame surface, and (3) the streakline pattern that results from the convection of the TiO_2 product downstream.

Both horizontal and vertical views of the flame were captured by the RMS technique. A camera positioned at right angles to a laser sheet passing through the center of the fuel jet provided a vertical view of the flame. The horizontal view was obtained by looking down at an angle of about 75° on a horizontal laser sheet. A pulsed Nd:YAG laser was used as a light source and was electronically triggered when the camera shutter was opened. Both the orange-colored blackbody radiation from the soot particles and the scattered green-colored laser light from TiO_2 particles (Mie scattering) were recorded on the film. The flame luminosity was integrated over the 2-ms camera shutter speed, while the Mie scattered light was captured during the 15-ns duration of the laser pulse. The luminous yellow flame surface appears to be 2D in the photographs, which is a 3D surface superimposed on the 2D photograph. Using the color discrimination between the luminous flame and the 2D Mie scattered light, one can obtain a clear interpretation of the flame photographs.

Mathematical Model

Formulation

A time-dependent axisymmetric model which solves for axial- and radial-momentum equations, continuity, enthalpy, and species conservation equations is used to simulate the reacting and nonreacting flow fields associated with coannulus jets. The body-force term due to the gravitational field is included in the axial-momentum equation. Density is obtained by solving the state equation while the pressure field at every time step is determined from pressure Poisson equations. Even though all the governing equations are solved in an uncoupled manner and iteratively, the species conservation equations are coupled through the source terms during the solution process to improve the stability of the algorithm.

In the present analysis of reacting flows a simple global-chemical-kinetics model involving propane, oxygen, water, carbon dioxide, and nitrogen is used. The stoichiometry follows

$$C_3H_8 + 5\,O_2 + N_2 \rightarrow 3\,CO_2 + 4\,H_2O + N_2$$

and the specific reaction rate is written in Arrhenius form. In order to represent very fast chemical kinetics, an activation energy of 1 kcal/mol and a pre-exponential constant of 1.0×10^{19} m^6/mol^2/s is used.

An orthogonal, staggered grid system with varying cell sizes in both the z and r directions is utilized. The momentum equations are integrated using an implicit QUICK-EST (quadratic upstream interpolation for convective kinematics with estimated streaming terms) numerical scheme (e.g., see Katta et al., 1994; Leonard, 1979) which is third-order accurate in both space and time and has a very low numerical diffusion error. The species equations, which have relatively large source terms, are integrated using a second-order upwind scheme. By rearranging the terms, the finite-difference form of each governing equation at all grid points is written as a system of algebraic equations which is then solved by using the alternative direction implicit (ADI) technique. The time increment, Δt, is determined from the stability constraint and maintained as a constant during the entire calculation. The pressure field at every time step is accurately calculated by simultaneously solving the system of algebraic pressure Poisson equations at all grid points using the LU (lower–upper) decomposition technique.

Temperature- and species-dependent thermodynamic and transport properties are used in present formulation. The enthalpy of each species is calculated from polynomial curve-fits, while the viscosity, thermal conductivity, and diffusion coefficients of the species are estimated from the Lennard–Jones potentials.

Boundary Conditions

The flow field considered in the present study has vortical structures of two different scales. Small-scale vortices develop on the fuel side of the flame (stoichiometric) surface along the shear layer of the fuel jet and larger-scale vortices are formed on the air side of the flame surface. The computational domain and the boundary conditions employed to capture these vortical structures are shown in Figure 1. The outer boundaries II and IV (cf. Figure 1) of the computational domain are located 40 and 15 nozzle diameters from the nozzle exit and the centerline, respectively, which are sufficiently far to minimize the propagation of disturbances into the region of interest. Flat velocity profiles are used at the fuel and air inflow boundaries. The outflow boundary in these flows is difficult to treat

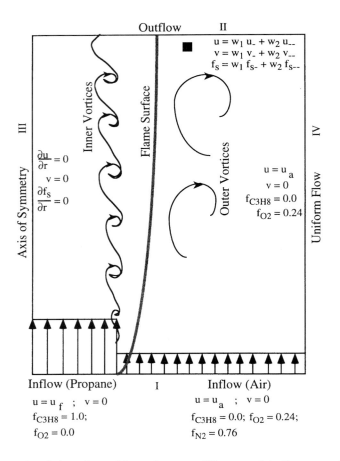

Figure 1 Computational domain and boundary conditions used in the numerical simulations.

because the flow leaving this boundary continuously evolves in time as the large outside and small inside vortices cross this boundary. A simple extrapolation procedure with weighted zero- and first-order terms is used to estimate the flow variables on the boundary. The weighting functions, w_1 and w_2 (cf. Figure 1), are selected by the trial-and-error approach; the main criterion used is that the vortices crossing the outflow boundary should leave smoothly without being distorted.

Results and Discussion

Vortical Structures

Several investigators have observed that jet diffusion flames possess longer potential cores and undergo slower transition to turbulence than the corresponding nonreacting cold jets (Yule et al., 1981; Eickhoff and Winandy, 1985; Coats and Zhao, 1988; Roquemore et al., 1989). These experiments indicate that the vortical structures in flames of low or transitional Reynolds numbers remain coherent for many nozzle diameters downstream. Typical flow fields representing cold and combusting jets are obtained using RMS technique and are shown in Figures 2(a) and 2(b), respectively. A long, straight tube with exit diameter of 11 mm was used in both experiments. Propane fuel was issued from the tube with a mass-averaged velocity of 2.0 m/s which corresponds to a Reynolds number of 3600 into a low-speed coannulus air flow. Since $TiCl_4$ was added only in the fuel jet of Figure 2(a), jet spreading and potential core may be identified from the outer and inner edges of

Chapter twelve: Dynamics of propane jet diffusion flames

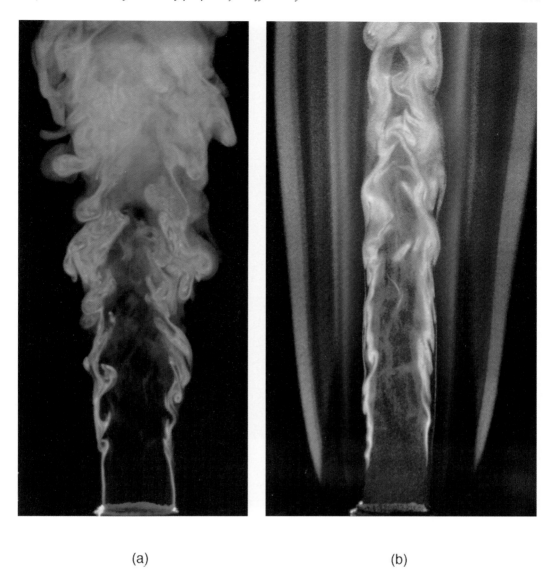

(a)   (b)

Figure 2 Propane jets emanating from 11-mm-diameter tube into low-speed coannulus air flow. Flow visualization is made using RMS laser-sheet-light technique. (a) Nonreacting (cold) flow at Re = 3626; (b) reacting (flame) flow at Re = 3988. Orange color represents the flame zone.

the green color, respectively. On the other hand, in the case of flame (Figure 2[b]), the potential core may still be identified in the same way; however, due to destruction of TiO_2 at flame temperatures, jet spreading may be obtained by locating the outer edge of the high-temperature or orange-color region. As expected, the spreading of fuel jet is much slower and potential core is longer in flame. However, the vortical structures in both the cold flow and flame are random and diffusive and do not exhibit coherence nature.

It is known that flow in a long, straight tube becomes turbulent for Reynolds numbers greater than 2000. Because the Reynolds number of the flows (3600) in Figure 2 has exceeded the transitional Reynolds number of the pipe flow, the fuel jet exiting from the central tube might have become turbulent which, in turn, effects the natural growth process of instabilities (or vortices) in the jet shear layer. To investigate the effect of combustion on jet spreading, the turbulence in the fuel flow at the burner exit was

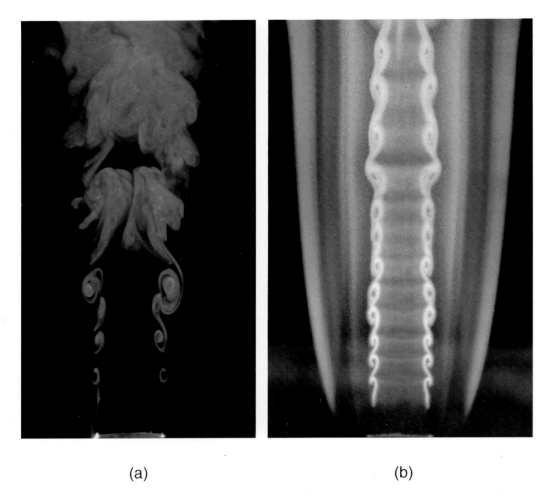

(a) (b)

Figure 3 Visualization of vortical structures using RMS technique in (a) cold flow at Re = 3988, and (b) flame at Re = 3988. Burner consists of 10-mm-diameter contoured nozzle and 150-mm-diameter annulus duct.

reduced significantly by replacing the long, straight tube with a larger diameter tube and a contoured nozzle. The structures of the cold and combusting flows obtained with contoured nozzle for a slightly different flow rate (Re = 4000) are shown in Figures 3(a) and 3(b), respectively. Velocity measurements at the nozzle exit indicate that the fluctuations are less than 1% of mean velocity. Similar to the observations made with tube flows (Figure 2) the contoured nozzle also results in less jet spreading and longer potential core for the flame (Figure 3[b]) compared to those for the cold flow (Figure 3[a]). However, in contrast, the contoured-nozzle flows yielded well-defined vortical structures in the shear layer. The coherence of vortices in cold flow is lost when vortex merging occurs at about two nozzle diameters downstream of the jet exit. The merged vortex pair remained intact while moving downstream for another three nozzle diameters before the vortex pair broke down into smaller fluid elements and the fluid elements exhibited a random and diffusive nature. However, in the case of combustion flow, no vortex merging was observed. The vortices remain coherent over a long distance and the vortex size also remains nearly unchanged.

In order to better understand the experimentally observed differences in the growth process of vortical structures in cold and combustion environments, these flows are numerically simulated using the time-dependent code discussed earlier. Numerical

Chapter twelve: Dynamics of propane jet diffusion flames

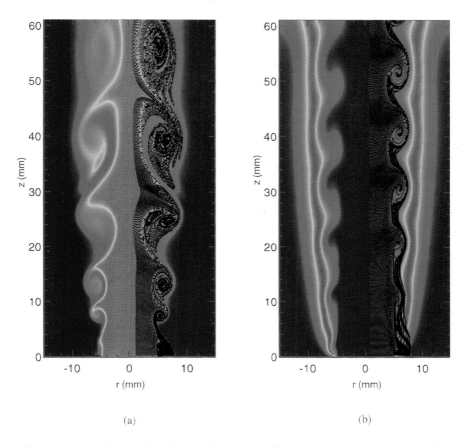

Figure 4 Numerical results obtained for cases shown in Figure 2 using axisymmetric, time-dependent model. (a) Iso-mass-fraction contours of propane in cold flow; (b) iso-temperature contours in flame. Instantaneous locations of particles injected at nozzle exit are superimposed with black dots on right half of each image.

simulations were carried out using flat velocity profiles at the exits of the nozzle and the annulus duct which correspond to the respective mass flow rates used in the experiment. As the shear-layer vortices in experiments manifest from the small disturbances that are normally present in the incoming jets, a small, forced perturbation is needed in the calculations for triggering the instabilities. Previous studies have indicated that perturbations provided in different variables such as velocity or pressure result in similar shear-layer structures. Therefore, in the present study, a perturbation of 0.3% in the axial velocity was introduced in the shear layer formed between the fuel and air jets. Inviscid stability analysis suggests that the most amplified frequency for the co-annular jet shear layer considered is about 400 Hz, using a Strouhal number of 0.2 with nozzle diameter and mass-averaged jet exit velocity being the length and velocity scales. In actual calculations, the perturbations introduced at this frequency did not travel beyond 0.08 m. The reason for this could be due to the viscosity of the fluid and the numerical dissipation inherent in the model which becomes increasingly significant at downstream locations where coarse meshes were employed. In order to account for these differences, the most amplified frequency in the simulations was determined by virtually repeating several calculations by varying the perturbation frequency. At 180 Hz the disturbances traveled farther downstream. Computed instantaneous cold flow and flame are shown in Figures 4(a) and 4(b), respectively. Cold flow is visualized in Figure 4(a) by plotting the fuel mass-fraction contours color coded between red and blue with these two limiting colors representing the pure fuel and no fuel, respectively. Instantaneous iso-temperature contours of the flame are shown in Figure 4(b) with red and blue colors representing maximum and room temperatures, respectively. In both figures, instantaneous locations of the massless particles that were introduced into the shear layer at exit of the nozzle are superimposed over the contour plots with black dots.

Similar to the observations made from experimental flow images, calculations also predict that the spreading rate of fuel jet is higher in the case of cold flow and the length of the potential core is longer in the flame case. Merging of two vortices in the cold flow is evident from the particle traces plot at about two nozzle diameters downstream of the exit ($z = 20$ mm). This merging location matches well with that observed in the experiments. The individual vortices that are involved in pairing kept their identity for another four nozzle diameters downstream (20 mm $< z <$ 60 mm) before the pair got involved in another merging process at about $z = 70$ mm. Even though experiments also show pattern for the convective motion of the paired vortices, results obtained with the present axisymmetric model become more skeptical soon after merging takes place. It is known that the azimuthal instabilities also become amplified during merging process of toroidal vortices and it results in three-dimensional flow. On the other hand, merging of vortices is not occurring in the flame case of Figure 4(b) and model predicted that these shear-layer vortices exhibit a high level of coherence and grow very slowly.

Larger view of the experimental flame in Figure 4(a) up to a height of about 180 mm is shown in Figure 5(a). The jet diffusion flame depicted in this figure exhibits two distinct vortical structures — one inside and the other outside the flame surface (orange-color region). While the smaller-scale inner vortices present in the jet shear layer lead to the development of turbulence in flames at higher jet velocities, the slowly moving outer vortices are associated with the low-frequency flame-flickering phenomenon first observed by Chamberlin and Rose (1928). As the jet shear layer is neither laminar nor turbulent, these flames are commonly referred to as transitional flames. The double-vortex structure of such transitional jet diffusion flame was first reported by Yule et al. (1981) and subsequently by Eickhoff (1982), Eickhoff and Winandy (1985), and Chen and Roquemore (1986).

The outer vortex structures of the flame in Figure 5 are the dominant characteristic of laminar and transitional jet diffusion flames. These slowly moving toroidal-vortex structures interact strongly with the flame and create outward bulges in the flame surface. The flickering appearance of the flame is the result of the upward convective motion of the flame bulge. Buckmaster and Peters (1986) were the first to recognize that buoyancy is responsible for the low-frequency instability associated with flame flicker. They argued that the natural convection of the flame and the forced convection of the jet are decoupled because the stoichiometric flame surface is located outside the shear layer of the jet; thus, the buoyancy-induced instability is nearly independent of jet characteristics such as fuel-jet exit velocity, nozzle diameter, and fuel type. By including buoyancy in the calculations they performed a linear instability analysis on a two-dimensional infinite candle flame. Their calculations resulted in a flicker frequency that was in good agreement with experiments. Recently, direct numerical simulations of buoyant jet flames have provided support and given additional insight into the findings of Buckmaster and Peters. By adopting a flame-sheet model, Davis et al. (1991) successfully simulated the dynamic structure of a 0.12-m/s buoyant propane jet diffusion flame established with a 22-mm-diameter tube. They showed that the flame is stationary and has no outside vortices when g was set to zero. In the presence of buoyancy, the outer vortex structures develop, and their upward convective motion is shown to be responsible for the 13-Hz flicker frequency associated with the flame bulge. Ellzey et al. (1991) successfully simulated a transitional hydrogen–nitrogen jet diffusion flame established by a 5-mm contoured nozzle having an exit velocity

Chapter twelve: Dynamics of propane jet diffusion flames

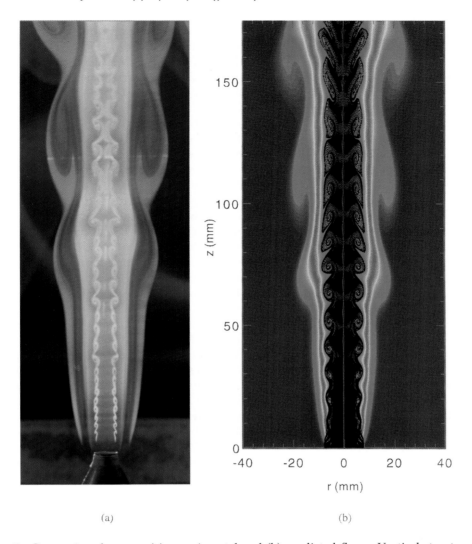

Figure 5 Comparison between (a) experimental and (b) predicted flame. Vortical structures and flame surface are visualized in experiment using RMS laser light and direct light, respectively. Particle traces and temperature contours show the computed inner and outer vortices, respectively. Red-color zone represents flame region.

of 10 m/s. They also showed that the convective motion of the outside vortex structures is responsible for flame flicker and that buoyancy is required to produce these structures.

The computed results obtained for the flame shown in Figure 5(a) are depicted in Figure 5(b). These results, in fact, represent a larger view of those shown in Figure 4(b). Because of the buoyancy term in the axial-momentum equation, vortices outside the flame surface (red-color region) developed as part of the solution. A comparison of experimental (Figure 5[a]) and computed (Figure 5[b]) flames indicates that the present calculations have captured the important details of a transitional flame.

A good agreement was obtained between the calculated and measured flame-flicker frequency. At 80 mm above the nozzle, a 15.8-Hz frequency was obtained from the computed temperature fluctuations, while the frequency was observed to be approximately 15 Hz in the experiments. The instantaneous iso-temperature color plot of the computed flame in Figure 5(b) is at a phase very near that of the experimental flame in

Figure 5(a). These instantaneous images show that the bulging and squeezing of the computed flame surface occur at heights that are slightly upstream of the corresponding processes on the experimental flame. It should also be noted that the computed flame has three outer vortices, while the experimental image shows only two and one half vortices. The flat velocity profile used in the calculations at the exit of the annulus duct could account for this difference. The rapidly converging nozzle in the experiment is believed to generate a significantly thick boundary layer (approximately 5 mm) for the annulus air flow. This reduction in velocity outside the flame surface could be responsible for the slightly lower frequency and larger-size vortices observed in the experiment.

The vortices inside transitional jet diffusion flame of Figure 5(a) maintained their coherence for long distances downstream. Comparison of Figures 5(a) and 5(b) shows that the model has captured this important feature. Even the coherence lengths in the calculated and measured flames are about the same. The inner (or shear-layer) structures of the experimental flame are weakened at $z = 165$ mm and completely dissipated at a height of approximately 210 mm. This weakening of the inner vortices appears to be associated with the third flame bulge. A change in the direction of rotation of the inner vortices may be seen in both the experimental and computed flames at a height of 150 mm and this process seems to be related to the second outward flame bulge.

The longer coherence length for the inner (or shear-layer) vortices in the case of combusting flow was conveniently explained by many authors (Eickhoff, 1982; Yule et al., 1981; Eickhoff and Winandy, 1985; Coats and Zhao, 1988) by the fact that the effective Reynolds number decreases due to the increase in viscosity. Chen et al. (1991) suggested that the vorticity destruction inside the flame due to the baroclincity and volumetric expansion laminarizes the flame. Recent calculations of Katta and Roquemore (1993) further indicate that the buoyant acceleration of the hot gases also plays a significant role in retarding the transition to turbulence and in sustaining the orderliness of the inner structures. The influence of viscosity and gravity on the inner vortices is illustrated in the numerical experiments of Katta and Roquemore (1993).

In a flame, combustion of fuel and oxidizer results in formation of products and heat release, which, in turn, effect the fluid dynamics of the flow by changing the density (or volumetric expansion), the transport properties, and the body forces due to buoyancy. When all these three changes are *absent*, the resulting flow will be identical to the cold flow as shown in Figure 4(a). Therefore, the noted differences in the growth of vortical structures in cold and combustion environments should be resulting from the changes in density, transport properties, and body forces. To investigate their role on vortex dynamics two additional calculations were made; one by not considering the changes in density and body forces and in another only the body forces were neglected. The effects of these three on the vortex dynamics are studied in the first two simulations. Color contours of instantaneous mixture fraction obtained for a flame with $g = 0$ and zero volumetric expansion are shown in Figure 6(a). The latter effect was simulated by maintaining the density of the fluid constant (incompressible) while allowing the viscosity, diffusion, and other transport properties of the flow to change with the calculated temperature. The solid circles represent the contour of the stoichiometric-mixture-fraction surface. Results obtained for the flame calculation with $g = 0$ are shown in the form of instantaneous-mixture-fraction color contours in Figure 6(b). Transport property calculations made in this simulation are identical to the ones made for the flame in Figure 6(a). Finally, the flame in Figure 6(c) is the mixture-fraction contour visualization of the flame in Figure 5(b). The green lines in this figure are the iso-temperature contours plotted only on the air side of the stoichiometric surface to visualize the outer vortical structures.

As described earlier, Figure 6(a) examines the effects of the temperature dependence of the viscosity and other diffusion coefficients, neglecting volumetric expansion. The stoichiometric surface moves away from the shear layer due to increased diffusion, and the

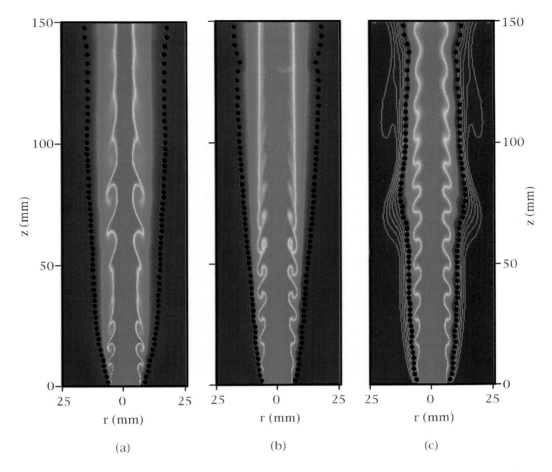

Figure 6 Iso-mass-fraction contours of propane obtained with different assumptions on combustion products. (a) Volumetric expansion is neglected; (b) gravitational force is neglected; (c) same as the flame shown in Figure 5(b).

viscous layer along the stoichiometric surface appears to damp the growth of the inner vortices (cf., Figures 4[a] and 6[a]). The first vortex-merging in this case takes place at a location between two and three nozzle diameters downstream. Since the viscous layer surrounding the shear layer acts as a shield to the shear-layer vortices, these vortices are stretched axially more than radially.

When heat release was also added to the calculations (Figure 6[b]), the inner vortices did not grow significantly and were completely dissipated at approximately 100 mm above the nozzle. The reason for this is that the volumetric expansion moves the stoichiometric surface closer to the shear layer, and viscous fluid near the flame surface is entrained into the vortices, leading to their dissipation. When buoyancy is added to the computation (Figure 6[c]), the flame surface moves even closer to the shear layer and one might expect the inner structures to decay more quickly than for the condition in Figure 6(b) because more viscous fluid is expected to be entrained into the vortices. Instead, the inner vortices in Figure 6(c) travel farther downstream with only a small growth rate. Therefore, a comparison of Figures 6(b) and 6(c) suggests that buoyancy is a major factor in maintaining the long coherence length of the inner vortices since it must overcome the dissipative effects of viscosity and volumetric expansion due to heat release.

Referring to Figure 4(b) it may be noted that the inner vortices are located in the blue region where temperature is low at a radial position between 4 and 7 mm. The high-temperature region (red color) is separated from the vortices. The acceleration of high-temperature fluid by buoyancy force entrains fluid from neighboring regions on both sides. Therefore, the amount of viscous fluid that is entrained into the vortices is reduced; the damping of vortices is reduced. The acceleration of fluid in the flame zone also reduces the velocity gradient in the shear layer which, in turn, reduces the entrainment of neighboring viscous fluid into the shear-layer vortices. In the case of cold jets, the axial-velocity gradient which is responsible for the formation of the shear-layer vortices is also responsible for their destruction as the higher velocity gradient leads to higher entrainment of the lower speed fluid into the vortices. Thus, the transitional jet flame and the cold jet differ in several important ways. In the flame case, fluid is entrained into the hot buoyant accelerated flame rather than into the inner vortices. The buoyant acceleration reduces the axial-velocity gradient of the jet shear layer, which reduces the tendency of fluid to be entrained into the inner vortices. In a cold jet significant entrainment of the low-speed fluid into the vortices leads to the loss of coherence of the vortices.

Vortical Structures of High-Speed Flows

As seen from Figure 2(a), a laminar jet flow becomes turbulent when the shear-layer vortices grow rapidly and involve in the merging process. When a diffusion flame forms in the neighborhood of the shear layer the growth of the vortices will be retarded and the jet flow tends to become laminar. However, at higher jet velocities, the growth of vortices in a flame is expected to overcome the dissipation caused by the combustion products. This phenomenon results in merging of vortices and a turbulent flame. The structure of a propane jet diffusion flame for different jet velocities was investigated experimentally and found that for jet Reynolds numbers greater than 9000 the flame becomes turbulent. The RMS images of the cold and combustion flows at a jet Reynolds number of 22,600 are shown in Figures 7(a) and 7(b), respectively.

As expected, the cold flow became turbulent very quickly after exiting from the nozzle. The vortices are quite small and do not exhibit any coherent behavior. A close look at lower and higher Reynolds number cold flows (Figures 2[a] and 7[a]) suggests the following important features of a turbulent jet flow: (1) at higher Reynolds number the jet shear layer becomes turbulent well before the entire jet does, (2) small-scale structures dominating the high-speed turbulent jet in which presence of large-scale structures is also evident, and (3) height of the potential core is more or less the same (about five nozzle diameters) for both the Reynolds-number cases.

The turbulent flame shown in Figure 7(b) is partially lifted and has an interesting pattern near the flame-base region. Due to the minor differences in the construction of the nozzle and outer duct, the jet flow is not perfectly symmetric about its axis. As the jet velocity of 9.8 m/s is very close to the lift off velocity for propane jet diffusion flames, the asymmetry in the burner geometry caused the flame to lift off on the left half (Figure 7[b]) and attached on the right half. Note, in the reacting flow experiments, only fuel jet was doped with $TiCl_4$ and the annulus air was dried to remove the moisture. The TiO_2 particles, which are marked by the green-colored scattered light, will form only when H_2O is present as a reaction product. Therefore, in the absence of flame (or combustion) in the left half of jet near the flame base, TiO_2 particle did not form and the shear-layer structures became invisible. As seen from the start of the green color, the lifted portion of the flame is anchored at an axial location of about two and a half nozzle diameters downstream of the nozzle exit. On the other hand, coherent (or organized) vortical structures have developed near the exit of the nozzle on the right half where the flame is attached to the nozzle. However, these vortices start losing coherence when merging of vortices begins to happen at about one nozzle diameter downstream of the exit and, subsequently, the shear layer became turbulent. Because of the cold conditions on the left half of the flame base, it is

Chapter twelve: Dynamics of propane jet diffusion flames 193

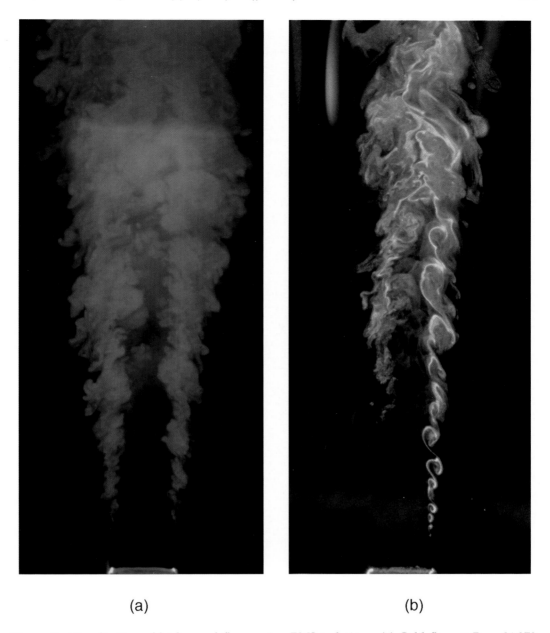

(a)   (b)

Figure 7 Visualization of high-speed flows using RMS technique. (a) Cold flow at Re = 21,270; (b) flame at Re = 22,600. Burner consists of 10-mm-diameter contoured nozzle and 150-mm-diameter annulus duct.

believed that the shear layer became turbulent quite early similar to that observed in a cold jet (Figure 7[a]) and overall the flow on the left half seems to be more turbulent than that on the right half.

Using the axisymmetric mathematical model described earlier, calculations were made for the high-speed reacting flow case shown in Figure 7(b). Unlike the fixed-frequency perturbation used for driving the shear layer of the low-speed jet (Figure 4[a]), a random noise was introduced in the shear layer of the high-speed jet to allow the calculations to respond to the most amplified frequency naturally. The random perturbation was applied to a region of about ten grid points within two-grid-point radius at the nozzle exit. No

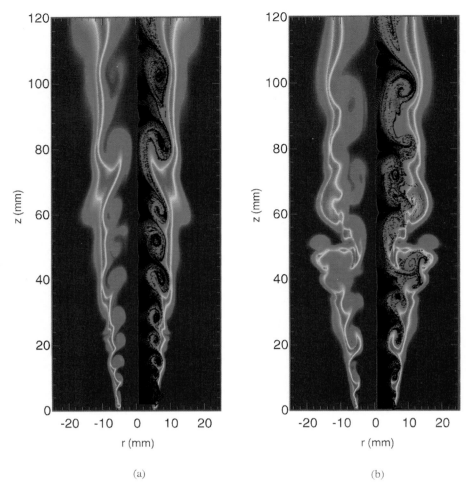

(a) (b)

Figure 8 Predictions made for flame shown in Figure 8(b) using two levels of perturbations. (a) 0.3%; (b) 0.6% of axial velocity. Instantaneous locations of particles superimposed over iso-temperature contours.

other perturbation was used, nor was another portion of the jet perturbed. Two calculations were made with different driving amplitudes to understand the effect of inlet turbulence on the flame structure. Instantaneous results for both cases are plotted in Figure 8 in the form of colored iso-temperature contours with locations of the particles superimposed with black dots. Figure 8(a) represents the results obtained for a driving amplitude of 0.3% of the axial velocity and Figure 8(b) represents those for a driving amplitude of 0.6%. For both cases, the perturbation of shear layer resulted in vortices with an initial crossing frequency of 1800 Hz. However, as seen from Figures 8(a) and 8(b), vortices grew rapidly and got involved in the merging process at about one nozzle diameter downstream of the exit and reduced the vortex crossing frequency to a value much lower than 1800 Hz. Even though the results obtained with an axisymmetric model might become skeptical once the vortices participate in a merging process, the data shown in Figure 8 depict some fundamental characteristics of a turbulent flame shown in Figure 7(b). When a small-magnitude perturbation (which represents a low-level initial turbulence) was used, the shapes of the shear-layer vortices (Figure 8[a]) were modified but not destroyed to an extent noted in the stronger-perturbation (which represents a high-level initial turbulence) case (Figure 8[b]). The overall structure of the flame in the former case matches with that observed in experiments conducted with contoured nozzles (Figure 7[b]), and the flame structure obtained with 0.6% driving resembles that obtained in tube experiments (not shown here). In neither case, the lift-off at the flame base was predicted as the model was constructed based on very fast chemical kinetics. In both cases of perturbations, puffs of fluid were found to eject out from the shear layer during the vortex-merging processes. Such fluid elements are experimentally observed by several investigators in the studies of turbulent jet flames (Takahashi and Goss, 1992) and are believed to play a significant role in stretching and creating localized holes on the flame surface.

Cross Sections of the Vortical Structures

To investigate the nature of the azimuthal instability in jet diffusion flames experiments were conducted by passing an additional laser sheet light across the flame surface. The RMS images obtained for a tube flow and for contoured-nozzle flows at different jet velocities are shown in Figure 9. These images depict both the longitudinal and azimuthal views of the shear-layer vortices. As seen from Figure 9(a), the azimuthal instabilities have grown to streamwise vortices within one nozzle diameter downstream of the exit. Similar to the toroidal (or ring) vortices that are visualized in the longitudinal section, these streamwise vortices also do not exhibit coherent behavior. Interestingly, the outer ring (green color) seen in the cross-sectional view is also asymmetric — indicating the three-dimensionality of the outer vortices that are developed due to gravitational force. As a very low speed (around 0.1 m/s) annulus flow was used in these experiments irrespective of whether a tube or contoured nozzle was used, the annulus air flow will be laminar. Therefore, it is speculated that the three-dimensionality observed for the outer vortices is a result of the rapid growth of the azimuthal instability of the inner vortices. This speculation may be confirmed by looking at the cross-sectional view of the flow structure obtained for the contoured nozzle at a moderate jet velocity (Figure 9[b]). In this case, the inner shear layer is laminar and the outer ring is quite symmetric even at an axial location of one and a half nozzle diameters downstream.

At transitional jet velocity a train of inner vortices has developed in the shear layer (Figure 9[c]). The cross-sectional view at an axial location of one and a half nozzle diameters downstream indicates that the inner shear layer lost its symmetry with the development of azimuthal instabilities; however, the outer flow remains to be symmetric. Since the inner vortices have not involved in any merging process, shear layer is dominated by the inner toroidal vortices that are nearly symmetric about the jet axis. Combining the observations made from the cross-sectional views of the inner- and outer-flow structures it may be argued that an axisymmetric mathematical model could be used for their prediction as long as the toroidal vortices keep their coherence and do not participate in a merging process. A three-dimensional model, however, should be used for the accurate prediction of flames in which merging of vortices takes place.

Conclusions

The dynamic characteristics of a propane jet diffusion flame for different jet velocities are investigated. To understand the role of combustion on the flame dynamics, cold (nonreacting) flows under identical conditions are also studied. First, the major differences are identified by dissecting the reacting and cold flows with RMS laser-sheet-lighting technique. Then, both of the reacting and cold flows are simulated by using a third-order accurate, time-dependent numerical code. Finally, numerical experiments are performed to gain further insight into the physical processes responsible for the observed dynamic characteristics.

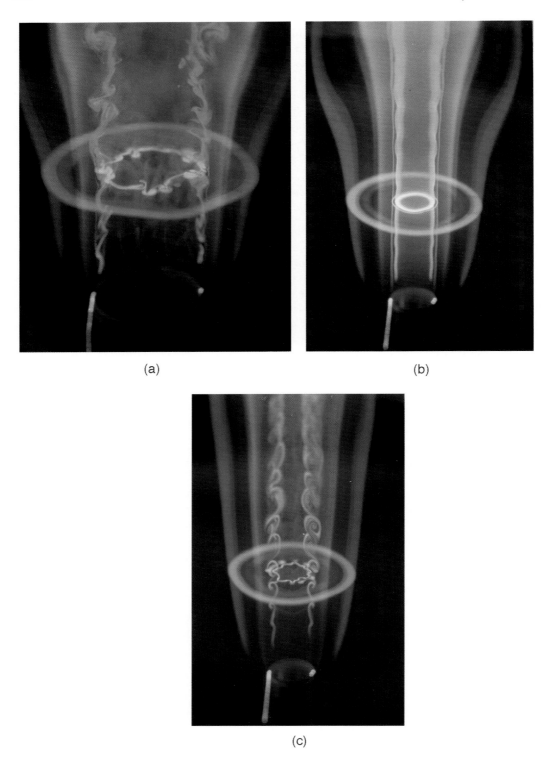

Figure 9 Visualization of streamwise vortices using a pair of perpendicular laser sheets. (a) Tube flow; (b) nozzle flow at low speed; (c) nozzle flow at transitional speed.

It was observed from the experiments conducted with straight tubes and contoured nozzles that the jet spreading rate is lower and potential core is longer for the jet diffusion flames compared to the nonreacting flows with identical flow conditions. Numerical simulations performed for these conditions also predicted the similar differences between the jet flame and the nonreacting jets. Numerical simulations showed that viscosity, volumetric expansion, and body force due to buoyancy associated with the combustion products are responsible for making the flames less turbulent. Particularly, buoyant acceleration of hot products restrains the vortices from entrainment, resulting in slowly growing vortices with longer coherence lengths. At higher jet velocities the shear-layer vortices inside the flame grow rapidly, involve in the merging process, and become turbulent similar to the scenario noted for the cold jets at lower jet velocities. The growth characteristics of the streamwise vortices resulting from the azimuthal instability of the shear layer are also investigated.

Acknowledgment

This work was supported in part by the Aerospace Sciences Division of Air Force Office of Scientific Research.

References

Bilger, R.W., Turbulent Jet Diffusion Flames, *Prog. Energy Combust. Sci.*, 1, 87, 1976.

Buckmaster, J. and Peters, N., The Infinite Candle and its Stability — A Paradigm for Flickering Diffusion Flames, in *21st Symp. (Int.) on Combust.*, Combustion Institute, Pittsburgh, 1986, 1829.

Chamberlin, D.S. and Rose, A., The Flicker of Luminous Flames, *Ind. Eng. Chem.*, 20, 1013, 1928.

Chen, L.-D. and Roquemore, W.M., Visualization of Jet Flames, *Combust. Flame*, 66, 81, 1986.

Chen, L.-D., Seaba, J.P., Roquemore, W.M., and Goss, L.P., Buoyant Diffusion Flames, in *22nd Symp. (Int.) on Combust.*, Combustion Institute, Pittsburgh, 1988, 677.

Chen, L.-D., Roquemore, W.M., Goss, L.P., and Vilimpoc, V., Vorticity Generation in Jet Diffusion Flames, *Combust. Sci. Technol.*, 77, 41, 1991.

Coats, C.M. and Zhao, H., Transition and Stability of Turbulent Jet Diffusion Flames, in *22nd Symp. (Int.) on Combust.*, Combustion Institute, Pittsburgh, 1988, 685.

Davis, R.W., Moore, E.F., Roquemore, W.M., Chen, L.-D., Vilimpoc, V., and Goss, L.P., Preliminary Results of a Numerical-Experimental Study of the Dynamic Structure of a Buoyant Jet Diffusion Flame, *Combust. Flame*, 83, 263, 1991.

Eickhoff, H., *Recent Contributions to Fluid Mechanics*, Springer-Verlag, New York, 1982.

Eickhoff, H., Turbulent Hydrocarbon Jet Flames, *Prog. Energy Combust. Sci.*, 8, 159, 1982.

Eickhoff, H. and Winandy, A., Visualization of Vortex Formation in Jet Diffusion Flames, *Combust. Flame*, 60, 99, 1985.

Ellzey, J.L., Laskey, K.J., and Oran, E.S., A Study of Confined Diffusion Flames, *Combust. Flame*, 84, 249, 1991.

Faeth, G.M. and Samuelson, G.S., Fast Reaction Nonpremixed Combustion, *Prog. Energy Combust. Sci.*, 12, 305, 1986.

Hawthorne, W.R., Weddell, D. S., and Hottel, H. C., Mixing and Combustion in Turbulent Gas Jets, in *3rd Symp. on Combustion, Flame and Explosion Phenomena*, Combustion Institute, Pittsburgh, 1949, 266.

Hottel, H.C. and Hawthorne, W.R., Diffusion in Laminar Flame Jets, in *3rd Symp. on Combustion, Flame and Explosion Phenomena*, Combustion Institute, Pittsburgh, 1949, 254.

Katta, V.R. and Roquemore, W.M., Role of Inner and Outer Structures in a Transitional Diffusion Flame, *Combust. Flame*, 92, 274, 1993.

Katta, V.R., Goss, L.P., and Roquemore, W.M., Effects of Nonunity Lewis Number and Finite-Rate Chemistry on the Dynamics of a Hydrogen-Air Jet Diffusion Flame, *Combust. Flame*, 96, 60, 1994.

Leonard, B.P., A Stable and Accurate Convective Modelling Procedure Based on Quadratic Upstream Interpolation, *Comput. Methods Appl. Mech. Eng.*, 19, 59, 1979.

Roquemore, W.M., Chen, L.-D., Goss, L.P., and Lynn, W.F., Structure of Jet Diffusion Flames, in *Turbulent Reactive Flows, Lecture Notes in Engineering, Vol. 40*, Borghi, R. and Murthy, S. N. B., Eds., Springer-Verlag, New York, 1989, 49.

Takahashi, F. and Goss, L. P., Near-Field Turbulent Structures and the Local Extinction of Jet Diffusion Flames, in *24th Symp. (Int.) on Combust.*, Combustion Institute, Pittsburgh, 1992, 351.

Wohl, K., Gazley, C., and Kapp, N.M., Diffusion Flames, in *3rd Symp. on Combustion, Flame and Explosion Phenomena*, Combustion Institute, Pittsburgh, 1949, 288.

Wohl, K., Kapp, N.M., and Gazley, C., The Stability of Open Flames, in *3rd Symp. on Combustion, Flame and Explosion Phenomena*, Combustion Institute, Pittsburgh, 1949, 3.

Yule, A.J., Chigier, N.A., Ralph, S., Boulderstone, R., and Ventura, J., Combustion-Transition Interaction in a Jet Flame, *AIAA J.*, 19(6), 752, 1981.

chapter thirteen

Application of Visualization Techniques for Studying the Internal Combustion Engine

Satoshi Yamazaki and Akinori Saitoh

Mechanical Division 1, Toyota Central R&D Labs. Inc., Nagakute, Nagakute-cyo, Aichi, Japan

> *Abstract*—A drastic reduction of fuel consumption and emissions is the urgent demand for internal combustion engines. To overcome this demand, a detailed understanding of the combustion process is necessary. Visualization of each process is the key to understanding. This paper attempts to show the various visualization techniques which have been developed for internal combustion engines and their results. The techniques were applied to the following phenomena: change of in-cylinder flow patterns, air/fuel mixing, flame propagation in diesel combustion, and changes in lubricant film thickness on the piston. From these results, much information has been obtained to help understand the complex interactions between the engine design parameters and combustion processes.

Introduction

Clean exhaust emissions and low fuel consumption have been continuously demanded of internal combustion engines. Such demands become more and more severe with the increasing awareness of global and urban atmospheric environments (Kawamura et al., 1988).

The demands will continue as long as the major power sources of passenger cars are internal combustion engines, that is, spark ignition and diesel engines. To overcome the current and future demands for cleaner emissions and low fuel consumption, more precise controls and design improvements of the engine system are necessary, based on the detailed understanding of the combustion processes.

Though the performance of internal combustion engines has been highly improved, the internal combustion process itself is still not clearly understood because of the complex interactions between the physical and chemical processes.

For more detailed and precise understanding of such complex processes, detailed numerical modeling is one of the helpful methods. Turbulent combustion modeling has been remarkably developed and is now partially applied to the engine design process.

On the other hand, observation and fine measurements of each internal combustion process are the key steps for the modeling and designing of the engine (Reeves et al., 1994).

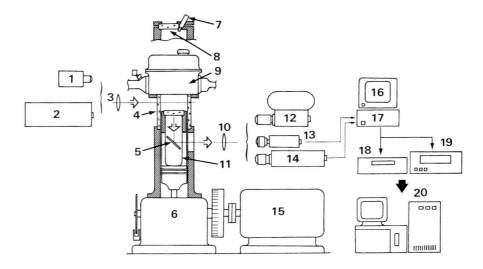

Figure 1 Schematic drawing of basic experimental arrangement for visualization of internal combustion engine.

This paper attempts to show some of the various visualization techniques which have been developed for internal combustion engine applications and their results.

Experimental Arrangement for Visualization of Internal Combustion Engine

A schematic drawing of the basic experimental arrangement for visualization of internal combustion engine is shown in Figure 1. A single-cylinder engine (built at Toyota Central R&D Labs., Inc.) is used as the base engine. It has a transparent piston and/or transparent cylinder. Moreover, a transparent head may be used for special purposes. The engine is connected to a dynamometer and operated in the motoring or firing condition. The base engine is capable of operating as a spark-ignited Otto-type engine (fueled by electronic-controlled fuel injector) or diesel-type engine (fueled by nozzle and high pressure fuel pump system).

Light sources are prepared for the specific purposes of visualization. A Xe flash lamp is used for illumination of instantaneous visualization for fuel spray shape, and a laser is used for inducement of fluorescence in fuel, mixture of fuel and air, and lubricating oil.

Several devices for imaging are used. They include a high-speed cine camera (which uses 16-mm cine film), a color CCD (charge coupled device) video camera with high-speed shuttering function, and an intensified CCD camera which has an image intensifier. The images of the CCD cameras are stored by VTR (video tape recorder) or laser disk. The stored images are then analyzed by an image analysis system.

Visualization of Each Process

Spray

Spray characteristics (spray formation, droplet diameters, droplet distribution, etc.) are usually measured in an injection chamber instead of an engine cylinder because of the difficulty in obtaining clear images. Especially, knowledge of the droplet distribution in a spray is important for understanding of the fuel/air mixture preparation.

Chapter thirteen: Visualization techniques for studying the internal combustion engine   201

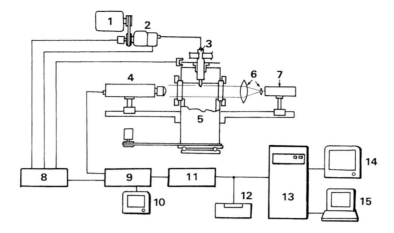

Figure 2 Measuring apparatus for laser light CT method. (From Kawamura, K., Saitoh, A., and Tanasawa, T., *Proc. 4th Int. Conf. on Liquid Atomization and Spray System*, 311, 1988.)

For the visualization of the cross section of sprays, a laser light computed tomography method has been developed (Kawamura et al., 1988). The principle is fundamentally the same as the X-ray CT method, which is widely used in medical diagnosis; however, for spray measurement a laser is used instead of X-ray.

The schematic diagram of the apparatus is shown in Figure 2. A 1-mW He–Ne laser is used for the light source, and this laser beam is expanded with lenses to form a parallel light flux. The light flux travels through the injection chamber. The transmitted light intensity is measured by the high-speed-shutter image-intensified video camera. The light source and the detector are rotated around a fixed nozzle, and thus the transmitted light intensity is measured from multidirections. There are 89 measurement points in the X-axis direction and 72 points in the θ direction. The distribution of transmitted light intensities at a certain time is obtained two-dimensionally in the X and Z axis directions by the camera. These data are digitized by an 8-bit A/D converter and input into the image processing system. Then, calculations are performed in the system, and the results are displayed on the color video monitor.

Figure 3 shows cross-sectional views of the excess air ratio in sprays with increase of fuel injection pressure. We can understand the air entrainment in the spray from these pictures.

In-Cylinder Flow

In-cylinder flow patterns such as swirl or tumble have been commonly used on recent spark ignition engines to increase turbulent combustion velocity, resulting in an extension of the lean limit of the engine (Reeves et al., 1994).

The particle tracking method is applied to an engine to investigate the detailed flow characteristics. A 50 × 1-mm Ar-ion laser light sheet is introduced into the central section of the cylinder through a transparent piston. Polymer microballoons (mean diameter = 40 µm) are added as a flow tracer to suction flow in the intake manifold. Mie scattered light from the tracer is observed by a high-speed video camera with image intensifier through a side window.

Figure 4 shows typical flow observation results around the spark plug of an engine with tumble intake ports. A vertical vortex flow which can be clearly seen at 80 BTDC was observed during the intake stroke to late in the compression stroke. However, this flow is

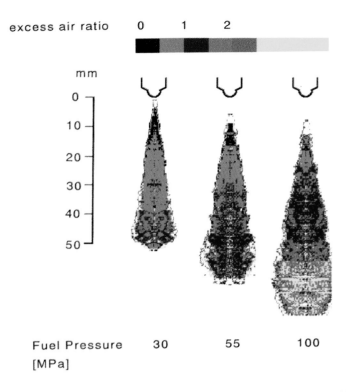

Figure 3 Cross-sectional views of the distribution of excess air ratio in sprays. (From Kawamura, K., Saitoh, A., and Tanasawa, T., *Proc. 4th Int. Conf. on Liquid Atomization and Spray System*, 311, 1988.)

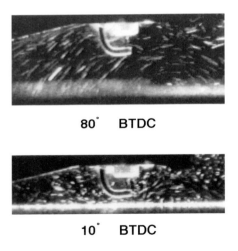

Figure 4 Particle tracking images.

extinguished rapidly as the piston approaches TDC, 10 BTDC, where dominant flow cannot be observed. Momentum energy of the tumble flow is transferred to turbulent energy when the vortex breaks down, leading to increases in the turbulent combustion velocity.

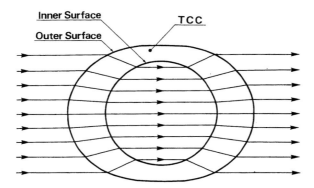

Figure 5 Cross-sectional view of the transparent collimating cylinder (TCC) and the path of light rays incident to the cylinder. (From Fujikawa, T., Ozasa, T., and Kozuka, K., SAE Paper 881632, 1988.)

Fuel–Air Mixture Preparation

Direct in-cylinder images of fuel–air mixture preparation and fuel distribution in firing engines are a great aid to engine design and optimization. Schlieren and laser-induced fluorescence (LIF) techniques to visualize these processes are reported in this section.

A transparent collimating cylinder (TCC) engine has been developed to obtain high quality schlieren photographs (Fujikawa et al., 1988; Fujikawa et al., 1994). Figure 5 shows the cross-sectional view of the TCC and the path of light rays incident to the cylinder. The inner surface has a cylindrical shape, while the outer surface has a specially designed shape to make light rays parallel at both the inside and outside of the cylinder. An in-house TCC made of quartz was mounted between the cylinder block and cylinder head with two gaskets.

Figure 6 shows the typical schlieren photograph of fuel spray during the injection period. In Figure 6, the fuel–air mixture is formed around the spark plug after the fuel spray impinges on the piston top surface.

LIF of commercial gasoline has conveniently been used for obtaining direct in-cylinder fuel distribution, because the addition of special dopants for LIF to the fuel is not necessary. In particular, planer LIF (PLIF), in which a laser sheet is used as an excitation source, is effective to obtain the cross-sectional image of fuel distribution, in contrast to the schlieren technique which gives us only the line-of-sight information (Hodres et al., 1991; Johansson et al., 1995). Because the absorption wavelength of liquid and vapor phase gasoline is shorter than 300 nm, a KrF excimer laser at 248 nm was used as the excitation source. Figure 7 shows the emission spectra of gasoline–air mixture gas excited by KrF excimer laser. LIF emissions extend from 270 to 400 nm with a maximum near 300 nm. The fluorescing components of gasoline are considered to be the aromatics (benzene, toluene, etc.). Because the LIF signal strength depends on the air–fuel ratio (A/F), it is possible to measure the quantitative fuel distribution from the image.

A schematic of the cylinder head is shown in Figure 8. Fuel distributions vary with the different operating condition of each fuel injector. The laser sheet of KrF excimer laser approximately 1 mm thick by 50 mm wide was split into two sheets by a half mirror. These laser sheets were introduced into the cylinder 2 mm below the spark plug. The bottom view technique was used. Mie and Rayleigh scattering was strongly attenuated with a long-pass glass filter. The fuel LIF was imaged by a quartz camera lens onto a CCD camera equipped with two-stage image intensifiers.

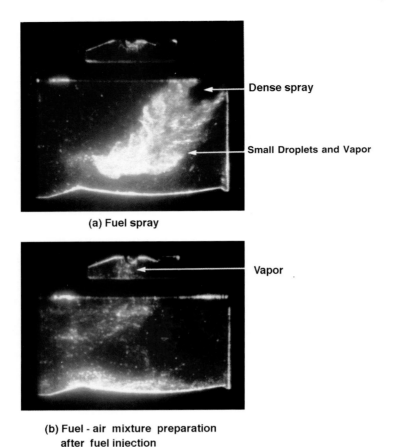

Figure 6 Schlieren photographs of fuel-air mixture preparation in the gasoline direct-injected engine. (From Fujikawa, T., Kataoka, M., and Katsumi, N., *Trans. Jpn. Soc. Mech. Eng. Ser. B* (in Japanese), 60-580, 4309, 1994.)

Figure 9 shows the typical LIF images of fuel distribution. In case 1, a nearly homogeneous fuel–air mixture is formed during intake and compression stroke. On the other hand, it is clear in case 2 that the liquid fuel droplets that are introduced from intake ports remain even late in the compression stroke. In case 3, fuel was supplied by only one injector (injector B in Figure 8). A stratified fuel distribution without fuel droplets can be seen during the intake stroke. This distribution was homogenized during the compression stroke.

LIF image of gasoline gives us the direct information about the fuel distribution.

Diesel Combustion Process

Combustion Zone: In the case of diesel combustion, nonluminous flames coexist with yellow luminous flames with very strong luminosity. Therefore, the image of a nonluminous flame appearing together with a luminous flame cannot be taken by ordinary photographing methods, due to the insufficient dynamic range of the sensitivity of film or video camera. In this case, a fuel additive is very effective. For example, copper oleate added to the fuel generates fairly intense luminosity in green due to flame reaction.

Images of the combustion zone (luminous flame zone) are shown in the photograph of Figure 10 (Nakakita et al., 1991). The observation field is indicated in Figure 10.

Chapter thirteen: Visualization techniques for studying the internal combustion engine 205

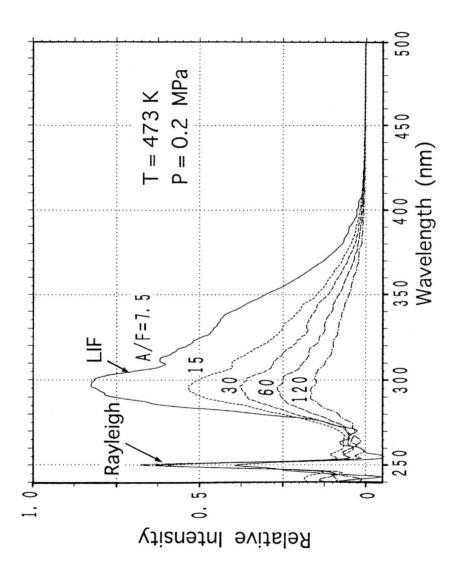

Figure 7 Emission spectra of gasoline–air mixture gas excited by KrF excimer laser at 248 nm.

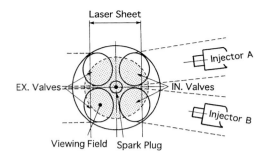

Figure 8 Schematic illustration of cylinder head used for the experiment.

Figure 10 shows a photograph of the nonluminous flame coexisting with luminous flame taken by using the 2 wt% copper oleate blended with diesel fuel. By image analysis based on the photograph, the ratio of nonluminous flame area to total flame area can be obtained, indicating the degree of fuel–air mixing.

Flame Temperature Distribution: Flame temperature distribution is obtained by the two-color pyrometry method (Matsui et al., 1978) using a two-color, high-speed shutter video camera system. An example of the flame temperature distribution is shown in Figure 11 (Kawamura et al., 1989). From this figure, low-temperature regions are seen to form near the fuel sprays injected in the flames and high-temperature regions appear in the developed flames downstream of the fuel sprays.

Not only flame temperature, but also KL values are simultaneously obtained by the two-color pyrometry method. As K is the monochromatic extinction coefficient and L is the flame thickness in the observed optical axis direction, the KL value is considered to be an index of soot density. An example of the KL value distribution is also shown in Figure 11. From this figure, it is clear that dense soot clouds are first formed in the regions of the spray-wall impinging points and next formed along the cavity wall and in the flame tips entering the top clearance zone.

Soot Formation Zone: There is another method for more direct visualization of soot clouds when a transparent combustion chamber with two windows at the opposite sides is available. The method is called back-illuminated photography (Nakakita et al., 1990). The observation results by this method are compared with those by other methods in Figure 12.

The basic information of the combustion field can be obtained from the schlieren photographs in Figure 12(a).

Figure 12 (b) shows photographs by the shadowgraph method, sometimes reported as the soot visualization method. Comparing the shadowgraph photograph with the schlieren photograph at TDC, however, it is clear that the shadowgraph method records shadows corresponding to fuel–air mixture. This indicates that this method is influenced even by the density gradients of fuel vapor. Accordingly, the shadowgraph images taken at 2 and 4° ATDC are generated not only by substances like soot, but also by the density gradients of gases. Therefore, it is difficult to visualize and detect in-cylinder soot clouds alone by the shadowgraph method.

Figure 12(c) shows photographs by the back-illuminated method. The mixture region at TDC is not recorded, and the shadow areas after 2° ATDC do not necessarily correspond to the combustion zone. It is clear, therefore, that this method is not influenced by the density gradients of gases, and that the shadows are due to substances which extinguish the incident light. However, not only soot but also fuel droplets block the light. Therefore,

Chapter thirteen: Visualization techniques for studying the internal combustion engine

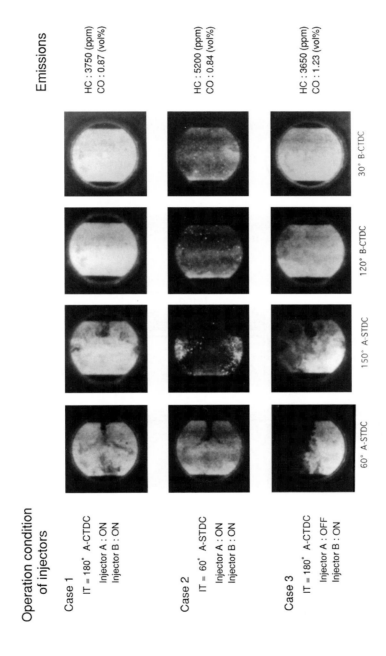

Figure 9 Gasoline LIF images for three operation conditions of injectors in Figure 8.

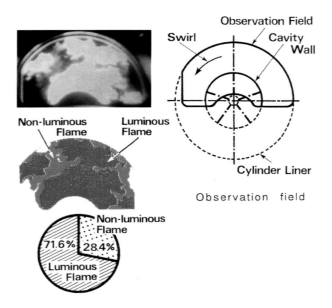

Figure 10 Example of a combustion photograph and its analyzed image (at 13 ATDC, inj. press.: 95 MPa). (From Nakakita, K., Miwa, K., Ohsawa, K., Takahashi, T., Watanabe, S., and Sami, H., *JSAE Rev.*, 12-1, 18, 1991.)

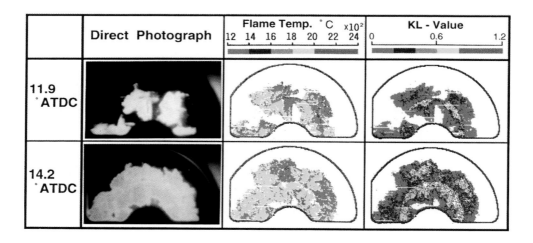

Figure 11 Distribution of flame temperature and KL value (inj. press.: 55 MPa).

the fuel spray area alone is recorded in Figure 12(d) of nonfiring-back illuminated photographs taken by injecting fuel in the combustion chamber filled up with nitrogen gas. From these results, the soot cloud is specified by subtracting the image of Figure 12(d) from that of Figure 12(c). In addition, the comparison of the back-illuminated photographs with the results by the transmissive light extinction measurement indicates that the back-illuminated images are the main soot clouds with light transmissivities less than about 70%.

Lubrication Around Pistons

In order to reduce friction loss and oil consumption around a piston, analysis of lubricant film behavior is essential (Ting, 1980). A single cylinder spark ignition engine for the

Chapter thirteen: Visualization techniques for studying the internal combustion engine 209

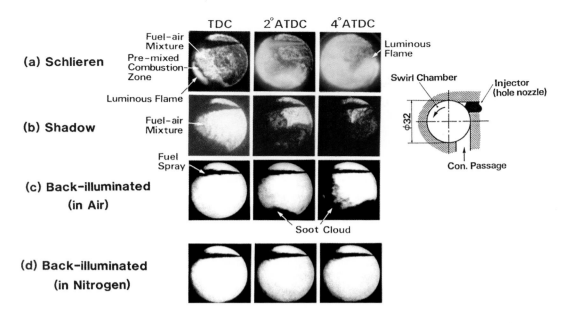

Figure 12 Observation results in swirl chamber by each visualization method. (From Nakakita, K., Nagaoka, M., Fujikawa, T., Ohsawa, K., and Yamaguchi, S., SAE Paper 902081, 1990.)

measurement has a cast iron cylinder liner having a pair of slit holes which are filled with sapphire glass, as shown in Figure 13.

The concept of the measurement is as follows (Inagaki et al., 1995): the combination of the Xe-flash lamp and the blue filter flashes blue light with a wavelength of 380 to 500 nm

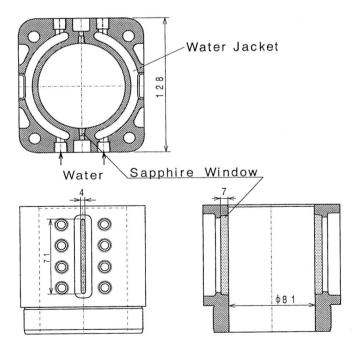

Figure 13 Cast iron cylinder with slit windows. (From Inagaki, H., Saitoh, A., Murakami, M., and Konomi, T., *SAE Paper 952346*.)

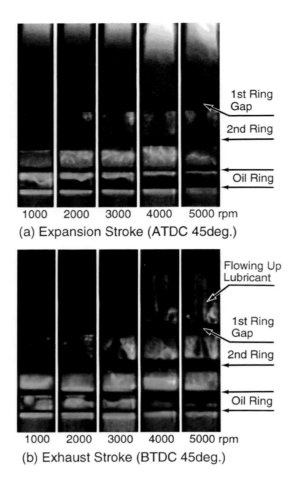

Figure 14 Lubricant behavior under full-load firing condition. (From Inagaki, H., Saitoh, A., Murakami, M., and Konomi, T., *SAE Paper 952346*.)

into the lubricant on the piston through the window at a period of less than 10 μs. A fluorescent dye coumarin-6 mixed in the lubricant is induced by the blue light, and consequently a yellowish-green fluorescence with a peak wavelength of about 510 nm is emitted with an intensity appropriate to the lubricant film thickness. The transient distribution of the fluorescence intensity at that period is recorded by the CCD video camera as a two-dimensional image. The reflected blue light is then excluded by the orange filter allowing only the light with a wavelength of above 500 nm to pass through. The resultant signals of the camera indicate quantitatively the distribution of the lubricant film thickness.

Figure 14 shows the effect of the engine speed on a full load firing condition. Under the high-speed operating condition, the lubricant flows upward toward the piston crown land.

Figure 15 shows the lubricant film thickness distribution on the piston skirt zone by using an all glass-cylinder during motoring. The results clearly show the area where the lubricant film thickness is less than 8 μm. From this analysis, we found where we have to actively supply lubricant to reduce the friction loss at the piston skirt zone.

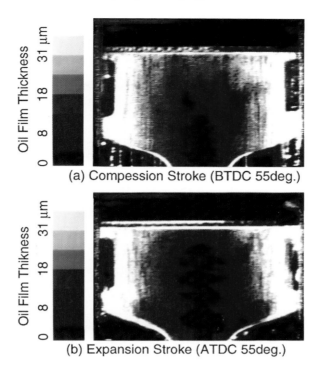

Figure 15 Lubricant film distribution at thrust-side piston skirt zone (engine speed: 1000 rpm). (From Inagaki, H., Saitoh, A., Murakami, M., and Konomi, T., *SAE Paper 952346*.)

Summary

A drastic reduction of emissions and fuel consumption is the urgent demand of internal combustion engines. To meet with this demand detailed understanding of the combustion processes is necessary.

Visualization of the processes is the key step for understanding. Though the current visualization techniques have some limitations to apply the internal combustion engines from the point of the quantitative measurements and also some modifications of the combustion chambers necessary to access the optical beams, these results give us much helpful information to understand the phenomena.

At the next step, two-dimensional LIF methods will be developed to measure the temperature and species distributions more quantitatively (Rakestraw et al., 1990; Vannobel et al., 1993). By coupling these results with numerical simulations, the combustion process will be more clearly understood.

Acknowledgments

I want to thank all my colleagues, Mr. T. Fujikawa, Dr. K. Akihama, Dr. K. Nakakita, and Mr. H. Inagaki, whose research works contributed significantly to this paper.

References

Fujikawa, T., Ozasa, T., and Kozuka, K., Development of Transparent Cylinder Engines for Schlieren Observation, SAE paper 881632, 1988.

Fujikawa, T., Kataoka, M., and Katsumi, N., Schlieren observation of in-cylinder processes with a transparent collimating cylinder, *Trans. Jpn. Soc. Mech. Eng. Ser. B,* (in Japanese), 60-580, 4309, 1994.

Hodres, J.T., Baritaud, T.A., and Heinze, T.A., Planar Liquid and Gas Fuel and Droplet Size Visualization In a DI Diesel Engine, SAE paper 910726, 1991.

Inagaki, H., Saitoh, A., Murakami, M., and Konomi, T., Development of two dimensional oil film thickness distribution measuring system, *SAE Paper 952346.*

Johansson, B., Neij, H., Alden, M., and Juhin, G., Investigation of the Influence of Mixture Preparation on Cyclic Variations in a SI-Engine, Using Laser Induced Fluorescence, SAE paper 950108, 1995.

Kawamura, K., Saitoh, A., and Tanasawa, T., Measurement of fuel concentration in spray by laser light computed tomography, in *Proc. 4th Int. Conf. on Liquid Atomization and Spray System,* Sendai, Japan, 1988, 311.

Kawamura, K., Saito, A., Yaegashi, T., and Iwashita, Y., Measurement of Flame Temperature Distribution in Engines by Using a Two-Color High-Speed Shutter TV Camera System, SAE paper, 890320, 1989.

Matsui, Y., Kamimoto, T., Matsuoka, S., and Oguri, A., Research on measurement of diesel flame temperature, *JSME,* 44, 228, 1978.

Nakakita, K., Nagaoka, M., Fujikawa, T., Ohsawa, K., and Yamaguchi, S., Photographic and Three-Dimensional Numerical Studies of Diesel Soot Formation Process, SAE paper 902081, 1990.

Nakakita, K., Miwa, K., Ohsawa, K., Takahashi, T., Watanabe, S., and Sami, H., Effects of high-pressure fuel injection on the combustion and exhaust emissions of a high-speed DI diesel engine, *JSAE Rev.,* 12-1, 18, 1991.

Rakestraw, D.J., Farrow, R.L., and Dreier, T., Two-dimensional image of OH in flames by degenerate four-wave mixing, *Opt. Lett.,* 15, 709, 1990.

Reeves, M., Garner, C.P., Dent, J.C., and Halliwell, N.A. Study of barrel swirl in a four-valve optical IC engine using particle image velocimetry, in Int. Symp. COMODIA 94, 1994.

Ting, L.L., Development of a laser fluorescence technique for measuring piston ring oil film thickness, *Trans. ASME Ser. F,* 165, 102, 1980.

Vannobel, F., Arnold, A., Buschmann, A., Sick, V., Wolfrum, J., Cousyn, B., and Decker, M., Simultaneous Imaging of Fuel and Hydroxyl Radicals in an In-Line Four Cylinder SI Engine, SAE paper 932696, 1993.

chapter fourteen

3D and 4D Visualization of Morphological and Functional Information from the Human Body Using Noninvasive Measurement Data

Naoki Suzuki and Akihiro Takatsu

Medical Engineering Laboratory, Department of Legal Medicine,
Jikei University School of Medicine, Tokyo, Japan

Preface

One major purpose of medical imaging is to observe the patient's internal condition without surgical intervention, something which cannot be done with the naked human eye. The history of medical imaging techniques began with the introduction of a method of diagnosis 100 years ago and this diagnostic goal has since been met in many ways. Invisible human structures were depicted as 2D images in the form of transparent or sectioned images. Diagnostic accuracy increased and provided an objective view with the application of these medical imaging modalities (Udupa, 1983; Barillot et al., 1985; Höhne and Bernstein, 1986; Robb et al., 1989; Robb and Hanson, 1990). The medical imaging techniques used in clinical medicine have become a necessary and indispensable element in current practice. It has also gradually become obvious, however, that medical 2D imagery in the form of transparent or sectioned images has certain limitations in viewing human structures. Advances in the capabilities of graphic computers and noninvasive imaging techniques have provided medicine with the advent of 3D or 4D imaging.

Furthermore, quantitative scientific visualization in 3D and 4D is not only useful for diagnosis, but also for more extensive applications such as surgical planning, treatment planning, and surgical support systems (Vannier et al., 1983; Linney et al., 1992; Dohi et al., 1990; Hashimoto et al., 1991; Joff et al., 1992). A significant number of 3D imaging techniques has already been utilized in routine clinical medicine, and these achievements further increase the possibility of other new medical techniques. Medical 3D imaging devices have led to the development of numerous techniques and software according to clinical demands and the numbers of such advances have been increasing rapidly in recent times.

However, we consider actual application to be only the tip of the iceberg in terms of the possibilities for high dimensional (3D and 4D) visualization of human structures. In actuality, there are problems that must be resolved to realize the hidden possibilities.

We have endeavored to develop a high-dimensional medical imaging system capable of visualizing and processing systemic structures in real time, or at high speeds close to real time, in order to observe or analyze morphological or functional features in an interactive way. We also applied a virtual reality technique as one means of handling high-dimensional medical imaging in an interactive way (Suzuki and Okamura, 1991; Suzuki et al., 1994; Suzuki, 1994; Suzuki and Takatsu, 1995).

Our aim was to identify the characteristics of high-dimensional imaging in medicine by referring to new approaches for visualizing the human body.

3D and 4D Visualization of the Living Human

Visualization techniques for invisible human structures were applied to diagnosis as long as 100 years ago. The X-ray photography developed by Röntgen was the starting point of this historical evolution. This technique was applied not only in an enormous number of clinical cases, but also led to other visualization techniques. X-ray computed tomography (CT) developed from the basis of X-ray photography in the 1960s and made it possible to see the living human in sectioned images. Advances in these noninvasive measurement techniques also led to ultrasonography and magnetic resonance imaging (MRI), and so on, in the 1970s. These visualization techniques fundamentally changed the algorithm method of diagnosis in medicine.

These techniques provide medical staff with internal human structures as a transparent image seen on X-ray or as a dissected image on CT or MRI. The idea of visualizing human structures as solids, as they exist in the body, was proposed when CT was first developed. Visualization of solid images of the body, which means reconstruction of a 3D image from sequential sectioned images, was initially very difficult. The processing speed and capacitance of computers were both inadequate at that time. In the 1980s the advances in computer technologies made it possible to process images containing enormous amounts of data at high speed. A 3D image of the human body, or a part of it, was gradually applied to clinical and basic sciences in the 1990s. Research on 3D image generation and applications to medicine increased during this period. However, it is said that these medical 3D image applications comprise only a small part of the present possibilities as described above. Furthermore, the human body is not a static system and shows dynamic changes in 3D structure as exemplified by the beating heart. Pulsation of the heart occurs via physical cardiac muscle dynamics and shows variable flow dynamics in the vascular system. Blood flows in the vascular system with a 3D structure including 4D vortex and turbulence. These complicated phenomena have now been observed precisely, whereas previously destructive measurements and total dynamics could only be imagined by physicians. 4D visualization of the human body, not only of cardiac dynamics but also locomotion and metabolism, would allow observation of these complicated and invisible phenomena. Only one method is considered to be available for visualizing and analyzing these complicated and time-sequential phenomena in the human body. A 4D image composed of time-sequential 3D images has made it possible to see such objects from the opposite direction of view at opposite time points by applying computer graphics techniques. However, medical 4D imaging techniques have not become popular in clinical practice due to the following problems. First, there is the difficulty of data collection in four dimensions. The second problem is the magnitude of the data explaining human structural dynamics. Human structures are composed of nongeometric shapes and multilayer and irregular spheres penetrated by branching vascularities. These structures require enormous amounts of data to be generated for their structures to be delineated in the computer space. Furthermore, sequential images require a higher quantity of data than single 3D images. The total quantity of 4D images can be expressed by summation of sequential 3D images in the process of showing individual stroke dynamics. Data collection at high-speed

Chapter fourteen: Visualization of morphological and functional information 215

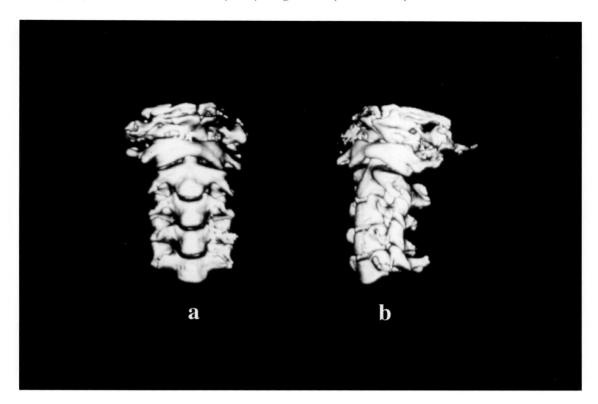

Figure 1 Reconstructed 3D image of the combination of cervical vertebrae. A luxation part is seen at the upper part of cervical vertebrae in the figure.

3D CT or 3D MRI, and so on, is now coming. The development of an ultrahigh-speed graphic computer allows the processing of these medical 4D data in large quantities in real time. In the next paragraph, I will explain the 3D and 4D visualization technique applied to human structure and function by referring to our experiments.

Visualization of Human Anatomy

3D Reconstruction of Hard Tissues

First, visualization of human morphology is divided into two categories. These are visualization of hard tissues and soft tissues. Hard tissues which include bone are easier to reconstruct and visualize as 3D images than soft tissues. The key point in this difference is the segmentation process. Specifically, segmentation of the region of interest (ROI) from volume data is very important for human structure visualization. We must select the ROI, such as specific bone or organ, in order to reconstruct it as a 3D image. If we neglect this process, the ROI will be buried in the 3D image of the most exterior component, the skin. In the case of CT examination, it is a very simple process to segment bones. That is, the X-ray absorption rate of bone tissue is extremely high as compared to other body tissues. Thus, we are able to segment bone tissues from surrounding soft tissues by simply thresholding the volume data. It is thus possible to display a complicated skeletal component such as the cervical vertebrae, as seen in Figure 1. The combination of cervical vertebrae, which have complicated shapes, is reconstructed as a 3D image automatically. A luxation part is seen at the upper part of the cervical vertebrae in the figure.

3D reconstruction of skeletal components was the first application of 3D imaging in the history of medicine and has been utilized in many plastic surgery cases (Robb and Hanson, 1990; Vannier et al., 1983). These hard tissue 3D images will increase the range of application if one problem can be overcome. This problem is that we cannot delineate a single bone from the skeletal system in an ordinal 3D reconstructed image.

Another processing technique is needed to separate individual bones from the skeletal system, as well as to allow movement so as to facilitate movement analysis. I will discuss these problems in the section "Visualization of Skeletal System Movements by 4D Imaging".

3D Reconstruction of Soft Tissues

It is very difficult to reconstruct a 3D image of soft tissues, such as organs and muscle, by both surface rendering and volume rendering methods. That is, we cannot omit the process of delineating contour lines of the ROI from surrounding tissues for 3D image reconstruction. This is the ROI segmentation process. Soft tissues, specifically organs, muscles, and vessels, cannot be segmented automatically with simple digital processing from volume data of either CT or MRI. This is because the value of the CT or MRI image does not reflect a specific part of the body. The same value, same graduation, and same layer condition are scattered in multiple areas throughout the body. Radiologists read and identify the shape and condition of the organ or vessels of interest based on their knowledge and experience. It is impossible to complete soft tissue segmentation using only digital processing. However, automatic reconstruction of an anatomical 3D image of the whole body cannot be realized without overcoming this problem. At present, there are research groups attempting to accomplish this full automatic segmentation and reconstruction of soft 3D tissue images by applying artificial intelligence, fractal theory, and so on. Now we will focus on one of our approaches, using Fuzzy theory for recognition and segmentation of the soft tissue of interest, from volume data (Matsushita et al., 1993).

First, a Fuzzy inference algorithm enables us to implement expertise as well as to integrate various image processing techniques, such as enhancement and sharpening. We have developed a Fuzzy inference board equipped with Fuzzy chips and installed it in a graphic work station. This enabled us to achieve higher precision, while reducing the processing time sufficiently for practical use. With this hardware we are endeavoring to segment the liver from the surrounding stomach, intestines, muscles, and other adjacent soft tissues. It is speculated that any soft tissue of interest will be segmentable with this hardware in the near future. In establishing these conditions, we will be able to observe or manipulate soft tissues in 3D space.

The images shown in Figures 2 and 3 are not a result of complete automatic image reconstruction of soft tissues with the aforementioned device. These images were made by applying an interactive process between computers and humans. However, these image creations will be fully automatic after the segmentation techniques have been realized.

Figure 2 shows images of the brain tissue reconstructed from sequential image CT data in 2-mm slices with a 2-mm pitch. Detailed morphology of the brain surface can be seen on the image. Figura 3a is an image of the heart reconstructed with data utilized under the former conditions. The internal structure of the heart, in Figures 3b and 3c, can be observed dissected in optional planes.

Human Atlas Database

Advances in processing speed and the capacity of the graphic work station allowed us to access a human structure 3D data base. Some research groups have started to establish

Chapter fourteen: Visualization of morphological and functional information 217

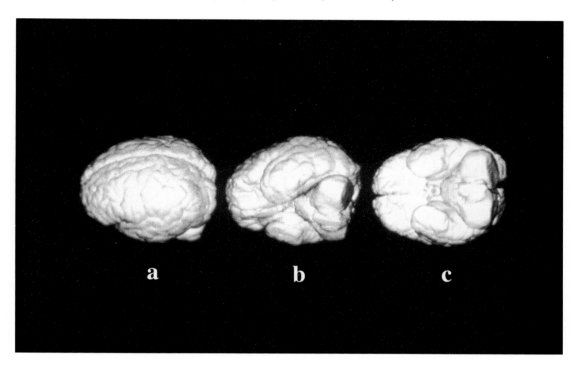

Figure 2 Images of the brain tissue reconstructed from sequential image CT data in 2-mm slices with a 2-mm pitch. Detailed morphology of the brain surface can be seen on the image.

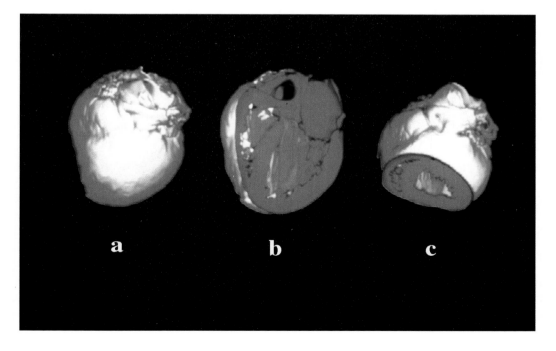

Figure 3 Images of the heart reconstructed from sequential image CT data in 2-mm slices with a 2-mm pitch. (a) Whole heart; (b and c) dissected in optional planes.

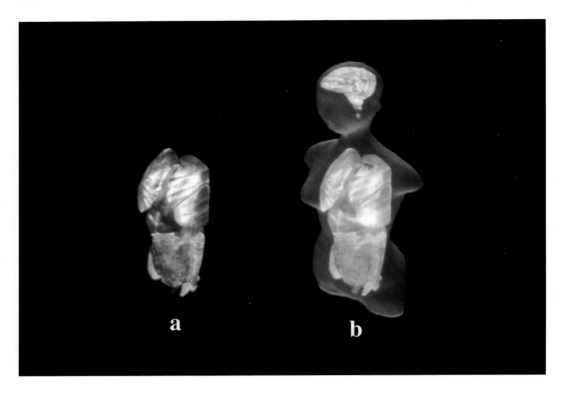

Figure 4 An image of the whole female body for human atlas database. Structural data are measured by MRI and color and texture information were collected from cadaver for anatomical education.

these human data bases named "Voxel man" or "Visible Human project" for such purposes (Höhne et al., 1992, Ackerman, 1991). We have also been constructing a 3D human atlas since 1986 using 3D reconstructed images from serial MRI. We made two 3D whole-body models, male and female, for this purpose (Suzuki et al., 1988). However, it will be a few years before we can manipulate these complicated images in real time. Figure 4 is an image of the whole female body. One characteristic of this system is that it contains detailed anatomical information on organ surfaces. It is impossible to obtain detailed information on organ surfaces, including color and texture, by noninvasive imaging methods. We have added this information to a 3D organ model obtained based on cadavers. We mapped organ textures by referring to the characteristic shape of each organ. These images can be rotated or enlarged on the overall image to 1024×1024 in real time.

Visualization of Blood Flow in the Heart Using Ultrasonography

It is possible to visualize not only morphology, but also functional aspects with 3D or 4D imaging. This paragraph will focus on the visualization of 4D blood flow distribution as an example of high-dimensional imaging.

The internal structure of the heart is very complicated and shape changes accompany the pulsation cycle. Blood flow dynamics in the heart are thus more difficult to understand than vessel flow dynamics. Clinicians have been using X-ray angiography, ultrasonography, and magnetic resonance imaging to assess these complicated blood flow patterns in the

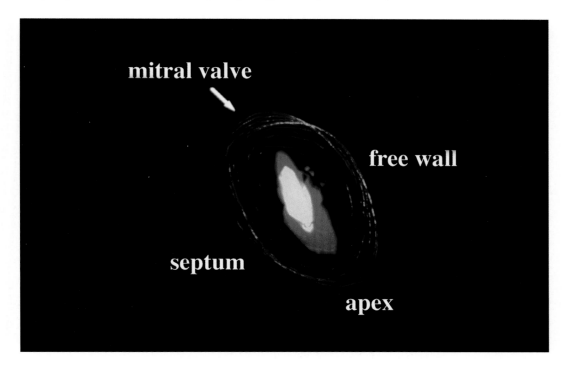

Figure 5 A reconstructed 3D blood flow image in the left ventricle of a normal subject. The mitral valve and names of other anatomical parts are seen in the image.

heart. Recently, in contrast to the conventional pulse Doppler method, circulatory morphology has come to be observed in 2D in association with the increasing prevalence of the color Doppler method. However, 3D and 4D analyses are required for precise observation of circulatory morphology, because blood flow in the intracardiac cavity has an essentially spatial dimension and changes within one cardiac cycle. Our group wanted to develop a system which would allow observation of blood flows in the heart comparable to flowing red ink in a glass of water. We designed a system to produce images of circulatory morphology as 3D images by using a computer graphic technique with blood flow images obtained by the 2D color Doppler method as original data (Suzuki et al., 1989). These 3D images can also be composed of 4D images using real-time image reconstruction. We examined the images for the purpose of applications such as the diagnosis of heart diseases and the evaluation of prosthetic heart valves. Original images of inflows into the left ventricle (LV) were measured as serial sectioned images. It is possible to produce 9 to 20 3D images in one cardiac cycle and these were displayed as 4D imaging using real-time image reconstruction. Moving blood flow images according to the phase can be rotated or dissected by the observer with a joystick or a dial. Figure 5 shows the reconstructed 3D blood flow image in the left ventricle of a normal subject. The mitral valve is represented in the upper column of the picture. Images of the inner wall of the left ventricle are depicted in the form of the wire-frame method to illustrate the internal details of flow architecture. The surface image in the wire-framed LV represents the shape and structure of inflow from the mitral valve toward the apex. The inflow surface was made transparent to show the region of the fastest flow rate within it. These reconstructed images were obtained in the early diastole phase in this case. The present method is partly characterized by observation of circulatory morphology from all directions by rotating the display of the reconstructed 3D images. Figure 6 shows the inflow changes in one cardiac cycle as a 4D image. Upper

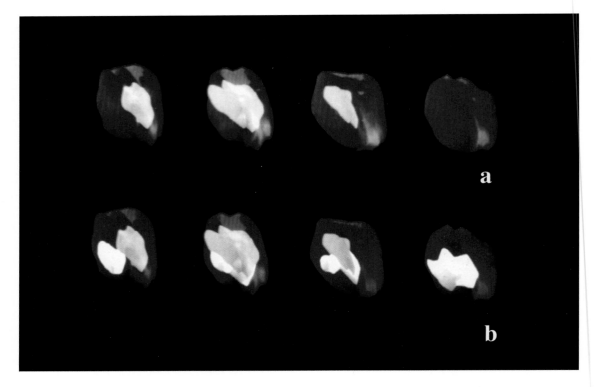

Figure 6 4D image of inflow changes in one cardiac cycle. Upper images display only inflows through the mitral valve and the flow areas to the apex have been added in the lower images.

images display only inflows through the mitral valve and the flow areas to the apex have been added in the lower images. The present method permits noninvasive 3D imaging of circulatory morphology. By rotating and transecting the image, 3D morphology including the orientation and structure of left ventricular inflow is assessed. The present study method was considered useful in diagnosis, assessing indications for surgery and evaluating the efficacy of surgery in various cardiac disorders including congenital heart diseases.

Visualization of Skeletal System Movements by 4D Imaging

It is very difficult to visualize relative skeletal system movements under free conditions with conventional medical imaging. X-ray photography provides the skeletal system image, but it is sometimes very hard to understand the 3D locomotion of articular movements or the movement of one bone in the skeletal system. X-ray movie has also been used for observations of skeletal systems, but this method is hampered by loss of information in the direction of depth of image. It is also difficult to comprehend the sectioned CT images. Clinicians must imagine these movements from ordinal X-ray photographs obtained in a few directions and naked eye views of the patient's movements involving the parts of interest when covered by skin. No present device used in noninvasive measurement for human diagnosis can meet this goal. Only 4D imaging applying computer graphics makes it possible to observe the relative movements of bones or the condition of joints. In this type of 4D observation, each bone is reconstructed from serial CT images and each location and movable range are labeled according to anatomical information and the recorded locomotion of patients. In this process, ordinal volume-rendered images of bones

Chapter fourteen: Visualization of morphological and functional information   221

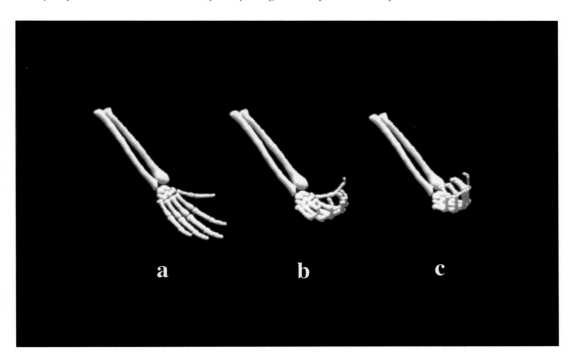

Figure 7 4D skeletal model of the human forearm from elbow to finger. The action of opening and closing the palm and relational conditions among individual bones can be observed as desired.

are difficult to use for this purpose. Each bone segment in the skeletal system should be identified and separated so that they can be manipulated as a 4D skeletal model.

Patients' movements are recorded using a location sensor and 3D skeletal images are manipulated with these location data. With these skeletal 4D images, the condition of one bone segment within the surrounding skeletal system and the complex function of joints are both visualized in detail in real time. Figure 7 shows a 4D skeletal model of the human forearm from elbow to finger. The action of opening and closing the palm and relational conditions among individual bones can be observed as desired. High-speed image reconstruction has made it possible to see these actions at a speed comparable to real human actions and also to see them in detail by magnifying the time scale.

These 4D analyses of human dynamics are expected to be achieved using real-time locomotion recording and virtual reality techniques.

Surgical Planning Using Virtual Reality

Virtual reality is based on an interface technique between humans and graphic computers and is most effective for obtaining practical information from subjects and relaying the user's decision in a natural way in a short time. Thus, this kind of interface appears suitable for a surgical planning system. That is, actual surgical manipulation or processes involving physical movements of the users' hands or real-time generation of 3D images of the operative field under the surgeons' fingers are now available with this interface. On the other hand, many cases in clinical medicine involve elastic and soft tissue materials such as organs, muscles, and vessels which are rather difficult to segment and reconstruct into 3D images. Our team attempted to develop a medical imaging system capable of handling soft tissues in real time, or at high speeds close to real-time to create a surgical planning system with reality, that is, real time manipulation allowing the user to perform virtual

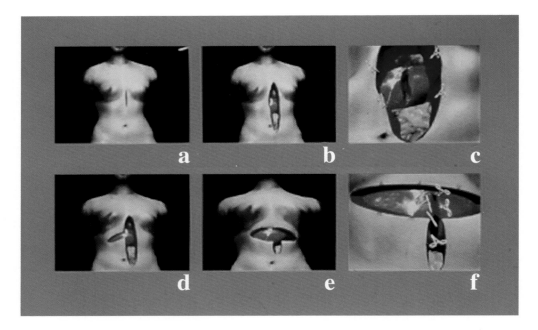

Figure 8 Procedure for making an incision from skin surface in the case of a hepatectomy lobectomy by a virtual surgery is seen in the images.

surgical operations on the image. We wanted to dissect the patient's organs in order to confirm the surgical approach or to optimize the procedure for the individual patient on the 3D image (Suzuki and Okamura, 1991; Suzuki et al., 1994; Suzuki, 1994; Suzuki and Takatsu, 1995). We undertook to develop a 3D simulation system for preoperative surgical planning of operations on organs. Targeted organs are visualized as 3D images using serial MRI or CT data from the patient, and the user is able to incise the organ surface and widen the incision line using surgical tools on a CRT with organ images.

This system is composed of a graphic work station and a pen-shaped or a glove-shaped input device with location sensor, that detects hand movements of the user. This is a man–machine interface system for visualizing procedure performed by the user, the surgeon. The patient's organ, including its internal structure, is reconstructed as a 3D image, and the portion to be incised is defined by the movements of the user's surgical knife or electric knife, as detected by the location sensor. Incision planes displayed on CRT are opened to the left and right against incision line, in a manner similar to an actual elastic organ, by calculating the direction, length, and depth of the incision. The incised part can be widened at the discretion of the user, and the sectioned vessels or other internal structures of the organ are also displayed on the surface of the incised plane. We used Onyx with Reality Engine^2 (Silicon Graphics, USA) as a main image processor. The user can stop bleeding from these vessels using an electrical knife or forceps under the same conditions as the surgical knife. Figure 8 shows the procedure used for making an incision from skin surface in the case of a hepatectomy lobectomy. Liver surface and inner vascular systems are reconstructed by MRI image sets in 5-mm slices with 5-mm pitch from the patient. The incised part is widened and the dissected organs are seen on the incised plane. Forces are also located at the dissected large vessels in images. Figure 9 also shows the procedure in the case of a hepatectomy lobectomy and the virtual environment was made more similar to the operating room. It was found that quantitatively accurate realistic images of the surgical procedure can be obtained using this system. The optimal surgical procedure can thus be designed for each individual case.

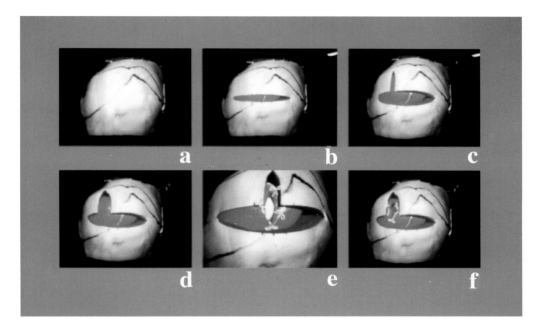

Figure 9 These images also show the procedure in the case of a hepatectomy lobectomy and the virtual environment was made more similar to an operating room.

Conclusion

Human structure is considered to exemplify the subjects which typically cannot be understood mechanistically with 2D imagery, but which can be dimensionally clarified in 3D or 4D. Human structure has a 4D component, and this phenomenon has imaging applications. This is clear when we observe the vascular system or the skeletal system as an example. Vascular system 3D structural changes in shape according to pulsation of the heart and blood flows in the vascular system also show a 4D distribution including complex vortex or turbulent flow patterns, as discussed in the section "Visualization of Blood Flow in the Heart Using Ultrasonography".

The visualization techniques applied to human structures, using our system developed in clinical experiments, have been described. There are two major elements which have advanced the visualization of human structure. The first is advances in the measurement techniques used in medicine. It has become possible to measure morphological and functional data in 3D or 4D, dimensions which have become easier and more rapid in recent years.

The technique of human visualization is based on the concept that the measurement should be done noninvasively. The noninvasive concept is similar in meaning to the nondestructive in the industrial sphere. Normal conditions of human structure and function can be visualized with data obtained from a living subject. 3D and 4D imaging techniques expand the capabilities of medical imaging and open the way to using images not only for diagnosis, but also for analysis of human structures and functions. Since ancient times, it has been an aspiration of medical practitioners to be able to observe patients' invisible internal structures just as we can see water in a glass. This ideal is becoming a reality thanks to advances in techniques in the field of medical visualization.

Future possibilities are expected to expand both advances in techniques used for noninvasive assessment of human structures and computer graphic techniques, including virtual reality, in the near future.

References

Ackerman, M.I. "Viewpoint: The Visible Human project." *J. Biocommun.*, 18, 14, 1991.

Barillot, C., Gibaud, B., Luo, L.M., Scarabin, J. M. "3-D representation of anatomic structures from CT examinations." *Proc. SPIE*, 602, 307–314, 1985.

Dohi, T., Ohta, Y., Suzuki, M., Chinzei, K., Horiuchi, T., Hashimoto, D., Tsuzuki, M. "Computer aided surgery system (CAS); development of surgical simulation and planning system with three dimensional graphic reconstruction." in 1st Conf. on Visualization in Biomedical Computing, IEEE, Tokyo, 1990, 458.

Hashimoto, D., Dohi, T., Tsuzuki, M., Horiuchi, T., Ohta, Y., Chinzei, K., Suzuki, M., Idezuki, Y. "Development of a computer-aided surgery system: three-dimensional graphic reconstruction for treatment of liver cancer." *Surgery*, 109, 589–596, 1991.

Heffernan, P.B., Robb, R.A. "A new method for shaded surface display of biological and medical images." *IEEE Trans. Med. Imag.*, MI-4, 26–38, 1985.

Höhne, K.H., Bernstein, R. "Shading 3-D images from CT using grey-level gradients." *IEEE Trans. Med. Imag.*, MI-15, 45–47, 1986.

Höhne, K.H. et al. "A 3D anatomical atlas based on a volume model." *IEEE Comput. Graphics Appl.*, 12, 72–78, 1992.

Joff, J.M., McDermott, P.J.C., Linney, A.D., Mosse, C.A., Harris, M. "Computer generated titanium cranioplasty: report of a new technique for repairing skull defects." *Br. J. Neurosurg.*, 6, 343–350, 1992.

Linney, A.D. et al. "The use of three dimension data on human body for diagnosis and surgical planning." *Rev. Neuroradiol.*, 5, 483–488, 1992.

Matsushita, S., Oyamada, H., Kusakabe, M., Suzuki, N. "Attempt to extract 3-D image of liver automatically out of abdominal MRI." *Proc. SPIE*, 1898, 803–808, 1993.

Robb, R.A., Hanson, D.P. "A software system for interactive and quantitative analysis of biomedical images," in *3D Imaging in Medicine*, Höhne, K.H. et al., Eds., NATO ASI Series, Vol. F60, 1990, 333–361.

Robb, R.A., Hanson, D.P., Karwoski, R.A., Larson, A.G., Workman, E.L., Stacy, M.C. "ANALYZE: A comprehensive, operator-interactive software package for multidimensional medical image display and analysis." *Comput. Med. Imag. Graph.*, 13, 433–454, 1989.

Suzuki, N. "Planning system and support system for surgery using virtual reality." in *Proc. 1st Int. Symp. on Computer Aided Surgery*, Tokyo, 1994, 20–21.

Suzuki, N., Okamura, T. "Support system for the field of surgical operation with 3D reconstructed image." *Dig. World Congr. Med. Phys. Biomed. Eng.*, 29(Suppl. Part 1), 451, 1991.

Suzuki, N., Takatsu, A. "Medical virtual reality system for surgical planning and surgical support." *J. Comput. Aided Surg.*, 1(2), 54–59, 1995.

Suzuki, N., Itou, M., Okamura, T. "Morphological reference system of human structure using computer graphics." *Phys. Med. Biol.*, 33(Suppl. 1), 90, 1988.

Suzuki, N., Okamura, T., Matsui, M., Arai, T. "Reconstruction of the three dimensional image of blood flows in the left ventricle." in *Proc. 11th Congr. of European Soc. of Cardiology*, Nece, 1989, 113.

Suzuki, N., Takatsu, A., Kita, K., Tanaka, T., Inaba, R., Fukui, K. "Development of a 3D image simulation system for organ and soft tissue operations." *Abstr. World Congr. Med. Phys. Biomed. Eng.*, 39a, 609, 1994.

Udupa, J.K. "Display of 3-D information in discrete 3-D scenes produced by computerized tomography." *Proc. IEEE*, 71, 420–431, 1983.

Vannier, M.W., Marsh, J.L., Warren, J.O. "Three-dimensional computer graphics for craniofacial surgical planning and evaluation." *Comput. Graphics*, 17, 263–273, 1983.

chapter fifteen

Giant Thermal Plume Generation Over Asphalt-Paved Highway

Manabu Kanda[1] and Mikio Hino[2]

[1]Dept. of Civil Engineering, Tokyo Institute of Technology, Tokyo, Japan
[2]Faculty of Policy Studies, Chuo University, Tokyo, Japan

Abstract—Large eddy simulations have been performed to investigate the characteristics of the convective boundary layer over the central part of Tokyo. Many thermal plumes generated over heated roads by solar radiation merge into giant walls of thermals. On the other hand, the downward sinking plume regions are formed around the green area surrounded by asphalt roads. Even for the real cases of nonuniform heating as treated in this paper, the nondimensional profiles of mean temperature and turbulent intensities agree well with laboratory experimental data over a uniformly heated surface and field observational data.

Introduction

During nocturnal time of clear-sky days, a thermally stable atmospheric layer is formed by cooling of the ground due to long-wave radiation. After sunrise and during daytime, violent thermal convection begins by heating the ground due to short-wave solar radiation to form a uniformly mixed layer with nearly constant potential temperature, called "convective mixing layer". Turbulent characteristics of the layer have been studied by several researchers, for example, a pioneering work by Willis and Deardorff (1974), an experimental study by Asaeda and Tamai (1982), and a numerical analysis by Schmidt and Schumann (1989).

Thermal conditions of real cities are nonetheless uniform. Thermal sources and sinks of various sizes are sporadically distributed; i.e., nets of roads paved with asphalt and tall concrete buildings in business zones are examples of heat sources, and parks, lakes, rivers, and green belts act as heat sinks.

Numerical simulations on urban climate were performed, for instance, by Kimura and Takahashi (1991) and Saito and Endo (1983), to elucidate the structure of meso-scale circulation over cities of horizontal scale 100 km. In these simulations, the sizes of numerical meshes are too coarse (about 2 km) to analyze the effect of the nonuniformity of surface thermal conditions. Recently, importance of the effects of geographical features on the urban climate has been pointed out from field observations. A belt of clouds formed above Main route No. 8 surrounding the Tokyo metropolitan area (the so-called "Kampachi cloud") is an example of such a phenomenon.

The aims of this study are to elucidate the effects of surface thermal condition on the turbulent structure of thermal convection developed over urban areas.

Methods and Conditions of Numerical Simulation

Fundamental Equations

The state variables treated in the atmospheric region are wind velocities (three components, u, v, w), atmospheric pressure (p), potential temperature (T), and humidity (q). The basic equations for these variables are grid-averaged. The LES turbulence model is applied in which turbulence production, termed "buoyancy", is taken into consideration. For flows through vegetation, the LES model of canopy layer is modeled after Kanda and Hino (1993; 1994). The SGS energy equation is solved simultaneously with the basic equations.

Computational Domain

The computational domain chosen for the study is the central part of Tokyo. A green area of rich vegetation in Emperor's Palace is surrounded by thermal sources such as a belt of metropolitan highway and tall buildings in official and business centers. The atmospheric computational domain of 4×4 km in the horizontal plane and of height 2.4 km in the vertical is divided into $45 \times 45 \times 45$ meshes, each mesh element on the ground being classified into either one of four categories of concrete, asphalt, vegetation, and water body. On the other hand, the computational domain of soils of 50 cm depth is divided into ten meshes, and the vegetation zones of height 2 m are also divided into ten.

Boundary and Initial Conditions

Periodic boundaries are assumed in the horizontal directions and a damping layer is installed on the upper boundary. On the lower boundary of the atmospheric domain, the Monin–Obukhov law is applied.

The computation starts from the meteorological condition of 10 a.m. (case 1); i.e., the initial thickness of mixing layer is 800 m, the potential temperature within the mixing layer is constant, and the potential temperature gradient of an upper stable layer is assumed constant (= 0.0035 K/m). The humidity of atmospheric layer is assumed constant (10 g/kg).

For vegetation zone, a new model named Neo SPAM proposed by the writers (Kanda and Hino, 1990) is applied. The vegetation model considers the uptake of soil water by root from the unsaturated zone of soil, the movement of water through trunks of trees and grass, the temperature decrease and moisture release by evapotranspiration of water vapor from leaves, and CO_2–O_2 exchange for photosynthesis. Decay of the insulation intensity through vegetation layers (considering leaf area density) is taken into account.

Numerical Computations

Computation for a uniform heating (case 2) has also been performed in order to compare the results with the nonuniform heating (case 1) for the real layout of thermal source and sink.

The computation was preceded by time step $\Delta t = 1.25$ s and the final time step was $N_{max} = 5000$, which is equivalent to real time of 1 h 45 min. The numerical computations were performed on a supercomputer NEC SX-2 (during September 1992 to 1993), CPU time for each run being about 10 h.

Chapter fifteen: Giant thermal plume generation over asphalt-paved highway

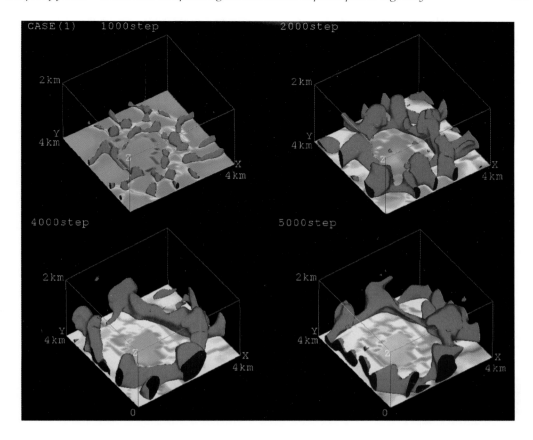

Figure 1 Sequence of the temporal development of plumes for the real situation (case 1).

Results and Discussions

General Features of Thermal Convection

Nonuniform Heating Case: Figure 1 shows a sequence of the computed results of the development of thermal plumes for the real situation of the central part of Tokyo metropolitan area — Otemachi and Emperor's Palace. Curved surfaces of plumes correspond to the iso-surface of the vertical wind velocity of w = 0.5 m/s.

Initially, small streets of thermal plumes develop along roads paved with asphalt where temperature rise is higher compared with concrete buildings and green area. As time step proceeds, these weak plumes merge each other similarly with the uniform heating case to form a wall of plumes over the metropolitan highway surrounding the heat sink of Emperor's Palace.

Figure 2 shows a perspective view at t = 6250 s of thermal plumes, the surface of which corresponds to the iso-surface of vertical velocity of w = 1.0 m/s. The warm-colored air mass represents the rising plume, while the cool-colored one is the descending plume. The central region of green area where no thermals are found acts as a large, cool island and forms a sinking flow region.

Uniform Heating Case: Figure 3 shows a sequence of the temporal development of thermal plumes for the uniform heating case. Curved surfaces of plumes correspond to the iso-surface of the vertical wind velocity of w = 0.5 m/s.

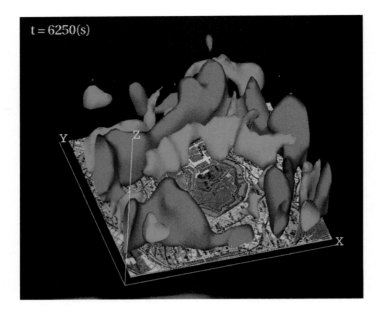

Figure 2 Bird's eye view of the rising (warm color) and descending (cool color) plumes at developed thermal stage, t = 6250 s.

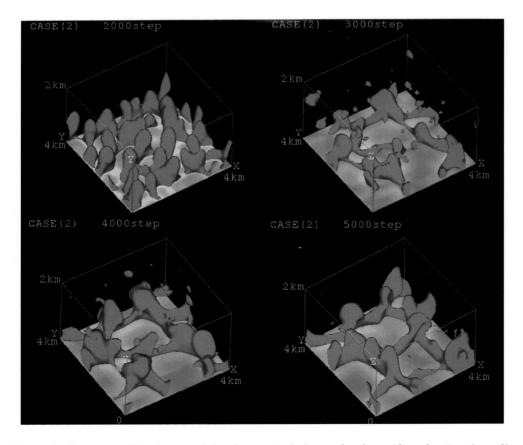

Figure 3 Sequence of the temporal development of plumes for the uniform heating (case 2).

Chapter fifteen: Giant thermal plume generation over asphalt-paved highway

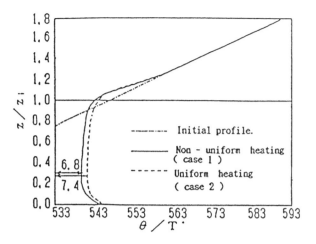

Figure 4 Profiles of the potential temperature.

After the beginning of heating, small thermal plumes are generated from randomly distributed sources. As time passes, they merge with each other to grow larger-scale thermals and finally to form distorted structures with a large scale, comparable to the nonuniform heating case.

Mean Profile of Potential Temperature

Mean quantities are defined as the spatial average in the horizontal plane and approximately equivalent to the ensemble average or temporal average of time series at a point (Deardorff, 1972). Turbulence characteristics are also defined as the spatial average in the horizontal plane of the deviations of a turbulent quantity from the above average.

In Figure 4, the profiles of the potential temperature after 4000 time steps for a uniform heating and the real situation are compared in nondimensional form. If the total heat flux supplied from the ground is assumed to be used for heating of air mass within the mixed layer, the temperature increase is, from conservation relation, $\Delta T/T_\star = 6.78$ for the thickness of the mixed layer, $z_i = 1060$ m. The values of $\Delta T/T_\star = 6.8$ (for case 1) and $\Delta T/T_\star = 7.4$ (for case 2) are only a little higher than the above estimation. The differences in the estimation and computational results on temperature increase in the mixed layer may suggest the additional heat supply from the upper warmer layer by thermal entrainment.

Figure 4 shows the so-called overshoot phenomenon, i.e., the temperature decrease at the bottom of the stable layer from the initial temperature profile.

Turbulence Intensity of Vertical Wind Velocity

The profile of the turbulent intensity of vertical velocity component of the present numerical experiments ($F_r = 0.12$) compared well with the data obtained under nearly the same stability conditions (Figure 5), i.e., the laboratory experiment ($F_r = 0.10$) by Willis and Deardorff (1974) and the field data measured by an air-borne instrument in an atmospheric mixing layer under the conditions of very low wind by Lenschow et al. (1980). Froude number F_r is defined as

$$F_r = w_\star/(Nz_i)$$

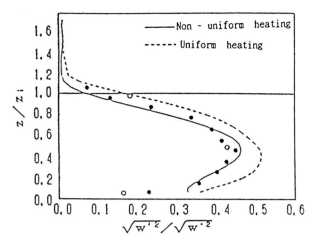

Figure 5 Profiles of the vertical turbulence intensity.

and

$$N = (\beta g \Gamma)^{1/2}$$

in which N means the Brunt-Baisala frequency, Γ is the gradient of potential temperature of the stable layer, and β means the thermal expansion coefficient.

Throughout the mixing layer, the vertical turbulence intensities for the uniform heating are 10% higher than those in the case of real nonuniform heating. The fact suggest the effective transfer of thermal energy given on the ground to air mass by rising and the mixing effect by buoyancy.

Turbulence Intensities of Horizontal Wind Velocity

The present results on the intensity of horizontal turbulent velocity (Figure 6) compared will with the experimental data by Willis and Deardorff (1974). Larger values of the horizontal turbulent velocity obtained in the field experiment by Lenschow et al. (1980) may reflect the increased turbulence produced by a light wind prevailing over the field.

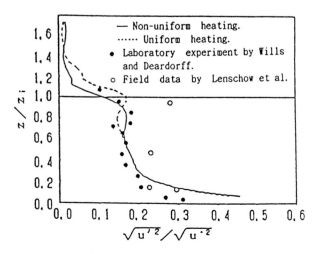

Figure 6 Profiles of the horizontal turbulence intensity.

Chapter fifteen: Giant thermal plume generation over asphalt-paved highway

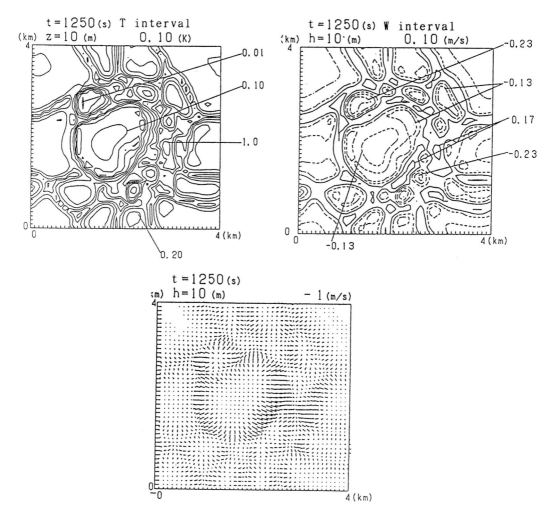

Figure 7 The horizontal patterns of convection at height z = 10 m at the initial stage of computation (t = 1250 s) for the real case (1). Contour plot of temperature (left), contour of the vertical velocity components (right), and velocity vectors (bottom).

Around the interface between the mixing layer and the upper stable layer, the turbulence for the uniform heating case exceeds that for the nonuniform case.

Horizontal and Vertical Patterns of Thermal Convection (Potential Temperature and Vertical Velocity)

The evolving processes of the thermal plumes are shown in Figures 7 through 10 for the real case (1) and Figures 11 through 14 for the uniform heating case (2) with regard to the velocity vector, vertical velocity component (w), and potential temperature. At the early stage from ground heating, many high temperature sources are randomly distributed, and no organized structure of plumes is recognized for the uniform heating case. On the other hand, asphalt paved roads and green and water areas around the palace are acting as thermal sources and sinks, respectively.

Although the distribution patterns of thermal plumes for the two cases of uniform and nonuniform heating are quite different, the mean characteristics of mean potential temperature and turbulent quantities are not so different as pointed out earlier.

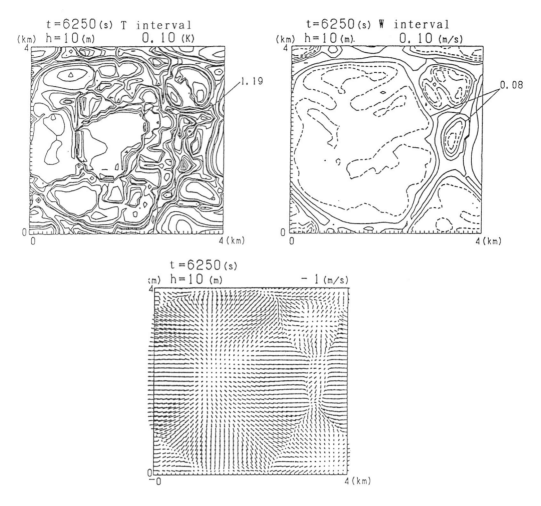

Figure 8 The horizontal patterns of temperature (left), velocity component (right), and velocity vector (bottom), at height z = 10 m and at the later stage of computation (t = 6250 s) of the real case (1).

These characteristics shown in the figures indicate the effectiveness of entrainment and mixing for a uniform heating, suggesting the mechanism of self-adjustment of thermal convection for the effective heat transport.

Acknowledgment

The present work was performed as a part of research project "Supercomputer and Society" (Chairman Prof. G. Yagawa, University of Tokyo, and Secretary General Mr. S. Kawasaki, Nikken Consultant Co.). CPU time of about 100 h on SX-2 was afforded us by Recruit Supercomputer Research Institute. The writers are very grateful for these persons and organizations.

Chapter fifteen: Giant thermal plume generation over asphalt-paved highway 233

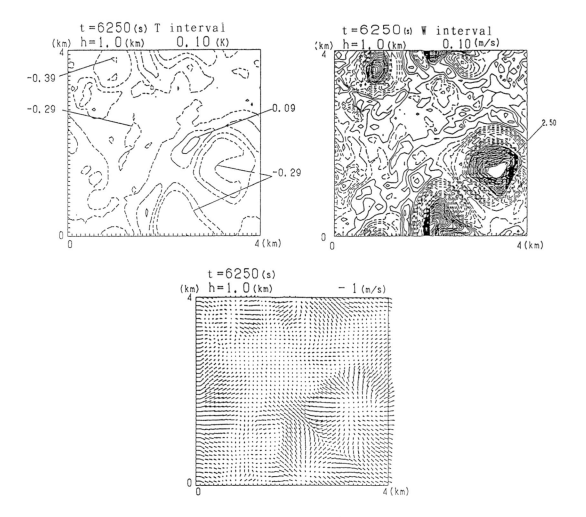

Figure 9 The horizontal patterns of temperature (left), velocity component (right), and velocity vector (bottom) at the boundary of the mixing layer; z = 1.0 km at the later stage of computation (t = 6250 s) for the real case (1).

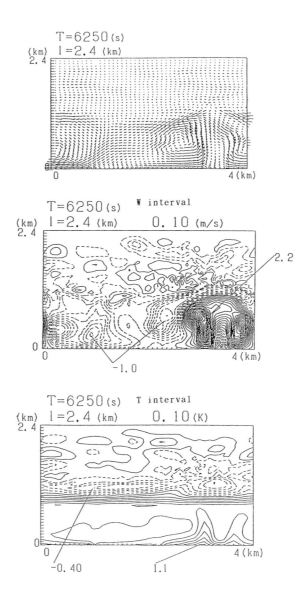

Figure 10 Vertical patterns of the velocity vector (upper), the contours of vertical velocity (center) and of potential temperature (lower), at the later stage of development (t = 6250 s) for the real case (1).

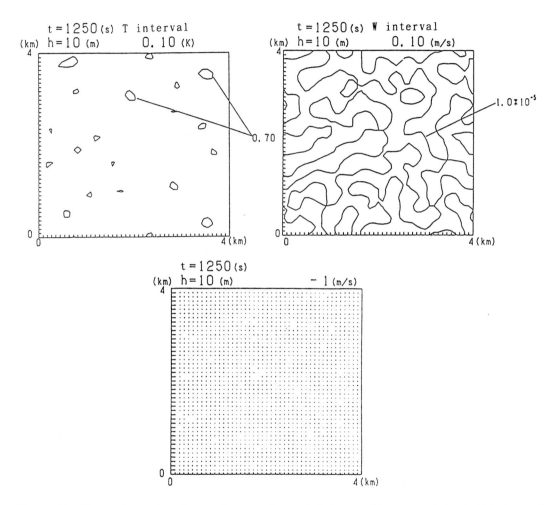

Figure 11 The horizontal patterns of convection at height z = 10 m and at the initial stage of computation (t = 1250 s) for the uniform heating case (2). Contour plot of temperature (left), contour of the vertical velocity components (right), and velocity vectors (bottom).

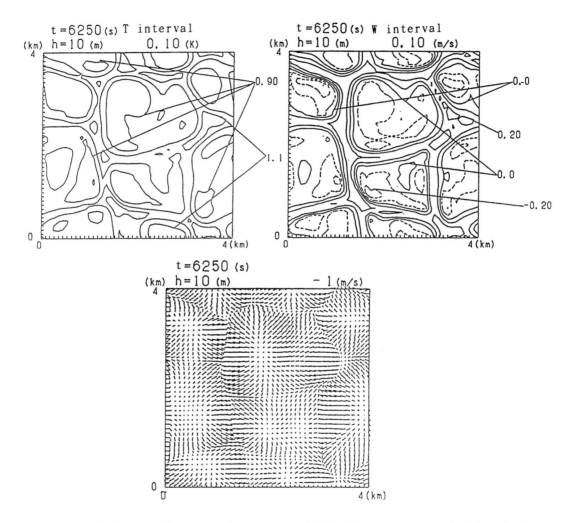

Figure 12 The horizontal patterns of temperature (left), velocity component (right), and velocity vector (bottom) at height z = 10 m and at the later stage of computation (t = 6250 s) of the uniform heating case (2).

Chapter fifteen: Giant thermal plume generation over asphalt-paved highway

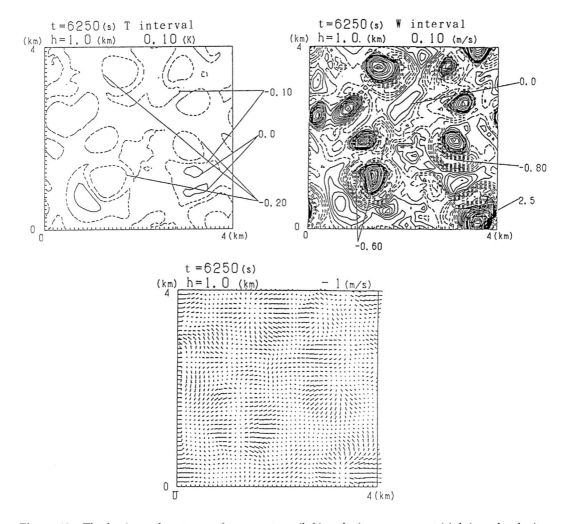

Figure 13 The horizontal patterns of temperature (left), velocity component (right), and velocity vector (bottom) at the boundary of the mixing layer; z = 1.0 km at the later stage of computation (t = 6250 s) for the uniform heating case (2).

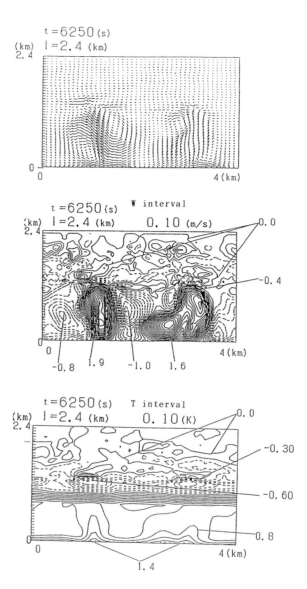

Figure 14 Vertical patterns of the velocity vector (upper) and the contour of potential temperature (lower) at the later stage of development (t = 6250 s) for the uniform heating case (2).

References

Asaeda, T. and Tamai, N., "On the characteristics of convection in a continuously stratified layer caused by thermal disturbances", *Proc. JSCE,* vol. 323, pp. 109–119, 1982.

Asada, T., Naganuma, T., Suzuki, M., and Shiozaki, I., "Thermal storage effect of pavement on urban warming", *Annu. J. Hydr. Eng. JSCE,* vol. 35, pp. 59–596, 1991.

Deardorff, J.W., "Numerical investigation of neutral and unstable planetary boundary layers", *J. Atmos. Sci.,* vol. 29, pp. 91–115, 1972.

Hino, M., Inagaki, S., and Kanda, M., "Effects of street-tree zone on the wind and thermal environment or urban street canyon", *Annu. J. Hydr. Eng. JSCE,* vol. 37, 1993.

Kanda, M. and Hino, M., "Numerical simulation of Soil-Plant-Air System. (1) Modeling of plant system", *J. Jpn. Soc. Hydrol. Water Resour.,* vol. 3, no. 3, pp. 37–46, 1990.

Kanda, M. and Hino, M., "LES modeling of the generation and development of convective cloud considering the interaction between air and soil systems", *Annu. J. Hydr. Eng. JSCE,* vol. 35, pp. 27–34, 1991.

Kanda, M. and Hino, M., "3-Dimensional turbulence structure of flow above and within the plant canopy", *Proc. 9th Symp. Turbulent Shear Flows,* pp. 12–1–1~6, 1993.

Kanda, M. and Hino, M., "Organized structures in developing turbulent flow within and above a plant canopy, using a Large Eddy Simulation", *Boundary-Layer Meterol.,* vol. 68, pp. 237–257, 1994.

Kanda, M. and Hino, M., "Thermal turbulence simulation by LES", 10th Symp. on Turbulent Shear Flows, Pennsylvania State University, 1995.

Kimura, F., "Heat flux on mixtures of different land-use surface; test of new parameterization scheme", *J. Meteor. Soc. Jpn.,* vol. 67, pp. 401–409, 1989.

Kimura, F. and Takahashi, S., "The effects of land-use and anthropogenic heating on the surface temperature in the Tokyo metropolitan area", *Atmos. Environ.,* vol. 25B, pp. 155–164, 1991.

Lenschow, D.H., Wyngaard, J.C., and Pennell, W.T., "Mean-field and second momentum budgets in a baroclinic convective boundary layer", *J. Atmos. Sci.,* vol. 37, pp. 1313–1326, 1980.

Sito, T. and Endo, K., "Three-dimensional simulation of urban heat islands", *Trans. JSME (B),* vol. 49, pp. 2035–2040, 1983.

Schmidt, H. and Schumann, U., "Coherent structure of the convective boundary layer derived from Large-Eddy Simulation", *J. Fluid Mech.,* vol. 200, pp. 511–562, 1989.

Willis, G.E. and Deardorff, J.W., "A laboratory model of the unstable planetary boundary layer", *J. Atmos. Sci.,* vol. 31, pp. 1297–1307, 1974.

chapter sixteen

Numerical Flow Visualization of Posuk-Chung, the Ninth-Century Remains of Poetry-Making Curved Water Channel in Korea

Keun-Shik Chang[1] and Eunbo Shim[2]

[1]Department of Aerospace Engineering, Korea Advanced Institute of Science and Technology, Taejon, Korea
[2]Department of Mechanical Engineering, Kum-oh National University of Technology, Kumi, Korea

> **Abstract**—The Curved Water Feast has been popular in history in the Far East Asian countries. The people in the ruling class of the old dynasties had enjoyed making poetry, sitting along the Curved Water Channel carrying a receptacle filled with liquor which was used for pleasant punishment when anyone of them failed to finish a poem. In this paper, the history of the Curved Water Channel is briefly reviewed for those channels existing in China, Japan, and Korea. Also, by taking advantage of the modern computational technique, the sinuous channel flow that had been enjoyed by the ancient people during the Curved Water Feast has been simulated with some approximations for the Korean Posuk-Chung, which is now completely defunct. Here, the interesting flow passage of the receptacle has also been numerically simulated. By this investigation, it has been possible to take a closer look into one of the elegant ways our ancestors had lived.

Introduction: History of the Curved Water Channels

The Curved Water Feast (曲水の宴) started in China in the 4th century has become very popular in the three Far East Asian countries, China, Korea, and Japan, until the 19th century before the last feudal dynasties collapsed. The Curved Water Feast used to be elegant merrymaking among the ruling class that included emperors, kings, feudal lords, the nobles, the gifted poets, and others. They sat around the curved water channel, composing short poetry in a limited time marked by arrival of the receptacle floating in the channel that carried liquor. Marvel and amusement crossed among the poets and the gallery depending on whether they succeeded or failed to finish a poem before the receptacle that had just arrived flowed away out of their reach; those who failed were pleasantly punished by the penalty of drinking liquor. To enhance the liveliness, the water channels were made very sinuous so that the receptacle could drift with prolonged time and possibly with interesting flow passage.

The origin of the Curved Water Feast goes back to the Jin Dynasty of ancient China when 42 poets including the famous calligrapher Wang Xizhi gathered on March 3rd, A.D. 353, at a curved water channel called Orchid Pavilion (蘭亭), in Shaoxing, Zhejian Province. They sat along the water channel by the bamboo trees to which the water was pulled from the nearby water pool, which was described quiet enough to reflect the surrounding mountains. In this Curved Water Feast, 11 poets composed two poems each, 15 poets composed one each, and the rest of them failed, to be punished by drinking three receptacles of liquor, respectively. Wang left the famous Orchid Pavilion Narration (蘭亭序) in the preface of those collected poems. A number of other curved water channels have been constructed since then, many of them having roofed shade above the channel to bear the name "pavilion". The channels were called in China by various poetic names such as Stamp Receptacle Pavilion (璽觴亭, implying perhaps a jade receptacle), Streaming Receptacle Pavilion (流杯亭), Prevailing Autumn Pavilion (沁秋亭), and Pliable Jade Pavilion (猗玗亭). Even more poetic names appeared in Qing Dyansty on the signboards of some particular pavilions (Zhao Luo, 1992): Curved-Creek Floating Flower (曲澗浮花) as brush-written by emperor Kang Xi (康熙帝) himself and Flowing Water Sound (流水音) as written by emperor Qian Long (乾隆帝), depicting the clear sound radiated from the small waterfalls built within the curved water channel. It is noteworthy that the custom of making excursion to the flowing waterside in early March of a new year had long been practised in China even before Wang Xizhi, and after as well, to welcome arrival of the spring and wash any sin and impurity. They believed it would prevent calamities which otherwise might inflict them.

The curved water channels are numerous in China, in addition to the Wang Xizhi's Flowing-Receptacle Curved Water (流觴曲水) the name of which was given by Wang Xizhi himself. Stone relics of a curved water channel were excavated in 1918 in the remains of Chong Fu Palace (崇福宮) of Song Dynasty, in Deng Feng Xian (登封縣) (Takeshima, 1970). The other curved water channels the authors have information on are all located in Beijing, most of them having been constructed in the Qing dynasty. It seems that the channels were very much enjoyed by the emperors of the Qing Dynasty. They are the Liu Bei Xue Lan Ting (流杯學蘭亭) in South Sea, the Ya Shang Ting (禊賞亭) in the Forbidden City, the Qin Qiu Ting (沁秋亭) near Shi Cha Hai (什刹海), the Liu Bei Ting (流杯亭) in Tan Zhe Si (潭柘寺), and the Zuo Shi Yan Liu (坐石焰流) in the ruin of Yuan Ming Yuan (圓明園) (Zhao Luo, 1992). There is even a contemporary Curved Water Channel constructed in bigger scale in the garden of Xiang Shan (香山) Hotel in Beijing. These channels are made of either jade or marbles; a Chinese literature called Ying Tsao Fa Shih (營造法式) written in Song Dynasty (10 to 13th century) describes the quantity of necessary stone and method of construction to build such a curved water channel. Figure 1 shows an example which appeared in Ying Tsao Fa Shih (Takeshima, 1970), showing a curved water channel with a decorative pattern on the stone surface. The perfectly symmetric shape of Ya Shang Ting in the Forbidden City, Figure 2, is not much deviated from this basic configuration, modified only by introducing one more inner loop in the center of the channel and by relocating the inlet and the outlet ducts on the opposite side.

In contrast, the Curved Water Channels in Japan are constructed as part of garden landscape and are, therefore, much larger than the Chinese counterparts; being very large, they have no pavilion above the channel. They are basically artificial creeks which simulate nature in reduced size as a typical Japanese garden does, constructed without carved granites or marbles. They used natural boulders, however, for the landscaping which were scattered alongside the channel; sometimes flat stones buried nearly to the ground level were used to make the waterway stone-paved.

The culture of the Chinese Curved Water Feast seems to have been propagated to Japan surprisingly early in the nation's history. There are a few documents which describe

Chapter sixteen: Numerical flow visualization of Posuk-Chung   243

Figure 1   A decorated Curved water channel in Ying Tsao Fa Shih (before 12th century).

Figure 2   Ya Shang Ting in the Forbidden City (Qing Dynasty).

that the Curved Water Feast occurred in the years A.D. 485, 726, 1007, and 1091 in Japan. The curved water channels in this country include the one at Jyonan-gu (城南宮) in Kyoto, the one in Gosho (御所) in Kyoto, the one in Kamikamo (上賀茂)-jinsha in Kyoto, the one in Mo-tsu-ji (毛越寺) in Hiraizumi (平泉), the relics in Imperial Palace of Nara,

Figure 3 The Curved Water Feast performed in Jyonan-gu, Kyoto, Japan (November 3, 1993). (By Courtesy of Prof. Y. Nakayama.)

the one in Dazaifu-Tenman-Gu, Ryuten (流店) of Korakuen Garden in Okayama, and finally the one at Isotenin (磯庭園) Garden in Kagoshima, Kyushu. These curved water channels have been constructed in different times of history ranging from the Nara period to the 18th century. These historic remains are cherished by the Japanese people; the ancient Curved Water Feast is even annually reproduced nowadays at some of the channels in the spring and in the fall before a large gallery; see Figure 3.

The culture of the Curved Water Feast has not skipped Korea in its course of propagation. It is not certain exactly when it was introduced to this country, but ancient history books (三國史記, 三國遺事) write that the 49th King Hunkang (憲康王) of unified Silla Dyansty (新羅) had made an excursion to the Posuk-Chung (鮑石亭) in March, A.D. 879. The name Posuk is origined from the shape of the stone channel resembling an abalone shell; see Figure 4. Unfortunately, it is the only curved water channel in Korea because a tragic accident occurred in A.D. 927 in which the 55th King Kyong-ae was slain during a Curved Water Feast in Posuk-Chung by the invading soldiers of the neighbor rival country, Hoobaikjae (後百濟). Posuk-Chung has been, since then, unjustifiably blamed for its ill fate and the culture of the Curved Water Feast itself has been totally eradicated in Korean history, in contrast to China and Japan where the same culture has prospered to outlive the Korean counterpart more than 1000 years.

Posuk-Chung is a curved water channel consisting of 63 finished granites which are assembled in the form of two side walls erected on the flat bottom plates. It measures 10.3 m in the long axis and 4.9 m in the short axis. It has a roughly symmetric oval shape in the main loop but is modified at two locations by a couple of highly sinuous curves. The channel width and depth vary somewhat along the waterway, but for the purpose of dimensional estimation one can say the channel is roughly 0.30 m wide and 0.22 m deep with 0.4 m-level difference between the inlet and the outlet. The channel currently does not

Chapter sixteen: Numerical flow visualization of Posuk-Chung   245

Figure 4 The defunct Posuk-Chung remains (constructed before A.D. 879).

have any pavilion despite its name and no record shows when it was built or lost. Having been defunct for long time, the channel nowadays has neither the water supply nor the drainage.

Looking into the fluid flows of the curved water channels by a modern computational technique could illuminate some of the fascination that had attracted the ancient people in the Far East Asian countries. There could be a reason, in addition to prolongation of time, why the channels were made highly curved without exception; sometimes they had very irregular parts added on to break the symmetry as seen in the Korean Posuk-Chung. Unlike other curved water channels as exemplified by a few Japanese channels which even nowadays witness annual reproduction of the ancient Curved Water Feast, the flow of Posuk-Chung is impossible to observe since the channel is completely defunct. In the present paper, computation of the channel flow has been performed to examine the flow structure of the Posuk-Chung. Also visualized is the flow passage of the receptacle. The results warrant invitation of those who have the slightest poetic mind to the vivid historic site of the ancient Curved Water Feast.

Computational Simulation

The curved water channel manifests elements of the river flow, meandering and braiding. One thing peculiar is that since it is relatively small, the surface tension does play a role, although the surface wave is not so significant. Some part of the curved water channel has been visualized by Nakayama et al. for the channel in Jyonan-gu, Kyoto, Japan, by means of white throat powder used in medicine. They observed stagnation, separated and reversed flows, and the vortices in the channel that make the flow passage of the receptacle very interesting. They also performed numerical investigation in two dimensions which provided some comparison with the experiment.

Previously, an experimental study using a model channel scaled down to 1/5.4 was made for the Korean Posuk-Chung (Chang, 1990) by the first author of this paper. It

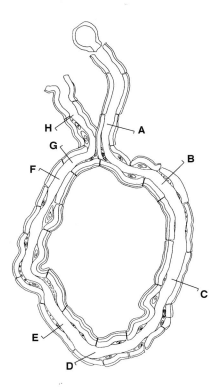

Figure 5 The separated vortex regions from the model experiment. (From Chang K.S., *Misoolcharyo*, Vol. 46, pp. 101–110 and 114–115, December 1990, National Central Museum, Korea [in Korean].)

revealed that the fluid flow was extremely complicated with many parameters to influence the flow. The surface flow visualized by the streaklines of the aluminum powder has exhibited separated flows and vortices in much larger scale than are normally expected in an open channel flow; see Figure 5. However, measurement of the important physical quantities such as the velocity, vorticity, pressure, turbulence, surface tension, surface waves, etc. in the whole channel flow has been practically impossible to perform. Even geometrical mapping of the flow passages of the model receptacle that depended on many flow parameters including the initial condition was far from being easy.

The physical flow throws much difficulty even to the computation. Do we have to use the moving mesh for the drifting receptacle? What happens to the mesh if the receptacle approaches the wall without any gap? The surface tension and surface waves will cause much difficulty if one insists on including them in the simulation. We will have to apply the free surface boundary condition at the boundary unknown *a priori* in this case.

We consider the three-dimensional incompressible Navier-Stokes equations,

$$\rho \frac{\partial \mathbf{u}}{\partial t} + \rho \mathbf{u} \cdot \nabla \mathbf{u} = -\nabla p + \rho \mathbf{g} + \mu \nabla^2 \mathbf{u} \qquad (1)$$

$$\nabla \cdot \mathbf{u} = 0 \qquad (2)$$

In the momentum and continuity equations above, $\mathbf{u}(u,v,w)$ is the velocity vector, p is the pressure, ρ is the density, $\mathbf{g}$ is gravitational acceleration, and μ is the coefficient of viscosity.

Considering complexity of the physical domain, we used the finite element method. The DOLFINS (drastically operator-lightened finite-element implicit Navier-Stokes) code, which has been developed by the authors and tested for a variety of problems, is used (Shim and Chang, 1994). In this method, the Galerkin finite element method is applied to Equations 1 and 2 to obtain the matrix equations,

$$M\frac{\partial U}{\partial t} + KU = CP + F \tag{3}$$

$$C^T U = 0 \tag{4}$$

where M is the mass matrix, K the convection plus diffusion matrix, C the gradient matrix, C^T the divergence matrix, and F the force vector. The DOLFINS code efficiently solves the above equations without iteration through one predictor step and two correction steps using the PISO (Issa, 1986) (pressure implicit with splitting of operators) analogy well known in the finite difference methods. The details of the method can be found in Shim and Chang (1994).

Flow in the Posuk-Chung Channel

Figure 6 represents the numerical model of the channel consisting of 10,780 surface grid points among 84,700 total points. Also shown inside of this channel is the experimental model previously used in Chang (1990). The flow enters the inlet plane with a predefined uniform velocity V_0 and leaves the outlet without flow gradient. The flow is assumed to have a constant depth and is shear-stress free (vanishing vertical velocity component) on the top surface. The present model thus excludes the gravity wave and surface tension effects. Further, the flow is driven by pressure gradient along a level channel. The viscous flow is subject to the no-slip condition on the channel wall.

The Reynolds number, $Re = \rho V_0 L/\mu$, of the real channel is believed to be of the order 10^5, while that used in the experiment (Chang, 1990) was in the range 500 to 3560, where L is the channel width in the inlet plane. It has been found in the experiment that the overall size of the separated flow region was rather insensitive to the Reynolds number.

The velocity vectors on the free surface obtained by computation are plotted in Figure 7 for Re = 500 in the laminar flow range. They present the separated flow regions well comparable with the vortex map in Figure 5, which was obtained by experiment for the turbulent flow at approximately $Re = 2 \times 10^3$. The channel flow was unsteady for higher Reynolds numbers in the computation, the separated flows being mixed intermittently with the main flow. The same phenomenon had also been observed and photograghed in the earlier experiment (Chang, 1990).

Enlargement of the channel flow is presented in Figures 8(a) to 8(d) for the four sections, AB, CD, EF, and GH, respectively, which are indicated in Figure 7. These sections correspond to the parts of the channel shown in Figure 5, where the channel appears most curved and irregular. The figures clearly show that the channel has surprisingly large separated flows in the parts having irregular width and large curvature. These flow patterns are well compared with the surface streaklines photographed in Chang (1990). The flow velocity near the bottom plane is retarded and modified, relative to the surface flow due to the drag exerted by the bottom surface. It will cause secondary flow in the curved part of the channel, producing three-dimensional flow structure. The braiding and meandering of a river is caused by the same flow mechanism.

Figure 9(a) represents distribution of velocity component normal to a channel cross section, C-plane marked in Figure 7. It shows that the stream velocity is low near the wall and high in the core region, with its center shifted toward the outer bank of the channel. The pressure distribution in the same C-plane given in Figure 9(b) indicates that pressure

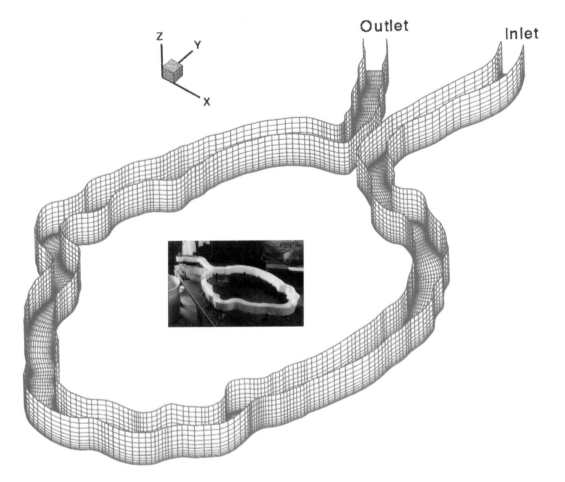

Figure 6 Numerical water channel with surface grid (inside of the channel is the experimental model used in Chang [1990]).

is highest on the bottom surface near the outer bank. The cross-stream velocity in Figure 9(c) clearly shows secondary flow in the C-plane of the channel.

The Trajectory of the Receptacle

The receptacle drifting on the free surface of the channel poses a very difficult problem classified as "moving boundary problems". The real receptacle has vertical dimension and therefore will have three-dimensional interaction with the channel flow as it drifts. Considering that the receptacle is relatively small in comparison with the channel width, we can assume that only the flow influences the passage of the receptacle but not vice versa. Then we can assume that the receptacle is represented by a particle mass by which the physical aspects of the receptacle motion is still preserved while difficulty if the grid singularity problem is avoided.

The Lagrangian equation of motion for the particle mass is

$$m_P \frac{dV_P}{dt} = F = \frac{1}{2} C_D \rho A \left| V_f - V_P \right| \left(V_f - V_P \right) \qquad (5)$$

Chapter sixteen: Numerical flow visualization of Posuk-Chung 249

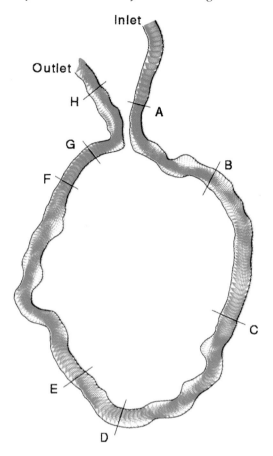

Figure 7 The velocity vectors on the top surface, Z = H.

where m_p is the particle mass, C_D is the drag coefficient, A is the cross-sectional area of the particle, and V_p and V_f are the velocities of the particle and the fluid, respectively.

Here we define the Stoker number regarding the particle mass as

$$St = \frac{m_p}{C_D \cdot \frac{1}{2}\rho A |V_f - V_p|} \quad (6)$$

which has dimension of time. We assume that the relative velocity $|V_f - V_p|$ can be replaced by a characteristic speed. The Stoker number then represents the inertia effect of the particle mass and is called "relaxation time". Equation 6 gives, after integrating for a small time step Δt,

$$V_p = V_f + (V_{p0} - V_f)\exp\left(-\frac{\Delta t}{St}\right) \quad (7)$$

where V_{p0} is the initial velocity of the particle mass. For a very small Stoker number the particle behaves nearly like fluid, but for a large Stoker number the inertia effect will give particle path significantly deviated from the streamlines of the surrounding fluid.

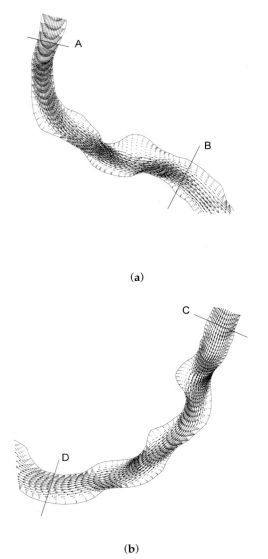

Figure 8 (a) Surface velocity vectors in section AB; (b) surface velocity vectors in section CD; (c) surface velocity vectors in section EF; (d) surface velocity vectors in section GH.

Suppose we can assume

$$\mu \sim O(10^{-3})\ N \cdot s/m^2$$

$$C_D \sim O(1)$$

$$A \sim O(10^{-1}L)\ m^2$$

$$|V_f - V_p| \sim O(V_0)\ m/s$$

$$m_p \sim O(10^{-2})\ kg$$

Then the Stoker number is in the order $O(10^2 Re^{-1})$. We computed the particle path for St = 0.3 while the initial position of the particle is varied.

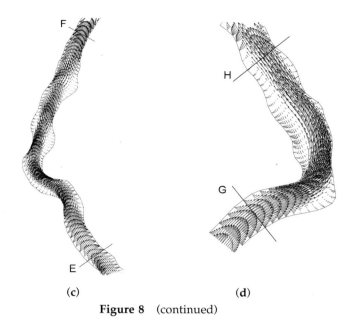

(c)   (d)

Figure 8 (continued)

    The receptacle trajectory is very much parameter dependent, creating certain unpredictability for arbitrary initial conditions. However, we can say that the particle mass would, in general, take a passage near the outer bank due to the centrifugal force and can be trapped in one of the major vortex regions. It many times escapes the vortex trap while circling in this region due to the inertia of mass and/or intermittent burst of the separated vortex. It then travels further downstream in the main stream and could be retrapped in a downstream isolated vortex. The particle mass, of course, can be drifted all the way to the exit without permanent trapping. This fact is entirely consistent with the observation made in the experiment of Chang (1990). Figure 10(a) shows that a receptacle started at the point P1 is once trapped in the location P2 before it escapes and is retrapped in a downstream location P3. A different picture appears in Figure 10(b) where the receptacle started at point P1 is permanently trapped at P2; when restarted at point P3, it flows along the outer bank all the way to the exit. In short, the time for arrival of the receptacle is many times prolonged in this way; the receptacle even had to be pushed out of the vortex trap for restarting when it was trapped for an excessive amount of time.

Conclusions

We have calculated the three-dimensional channel flow of Posuk-Chung in Korea used in the ancient time for the Curved Water Feast. Since the channel is defunct now, this numerical simulation seems to be the only realistic solution to the problem when detailed flow structure is desired. Despite its own approximations such as neglecting the surface tension, free surface effect, turbulence, and slope along the channel, in addition to the fact that the receptacle is replaced by a particle mass, we were able to identify the flow structure, separated vortex regions, and receptacle trajectories very similar to those qualitatively observed in the earlier model experiment. The passage of the receptacle suggests certain unpredictability since it depends on the unpredefined conditions such as stream speed, weight of receptacle, and starting position of the receptacle. This uncertainty would certainly be one of the interesting aspects of the Curved Water Feast.

    By this research, we had opportunity to review the literature in China, Japan, and Korea in a comparative manner, and study some new fluid dynamic aspects of the Posuk-Chung; some of its findings could be common among the curved water channels scattered

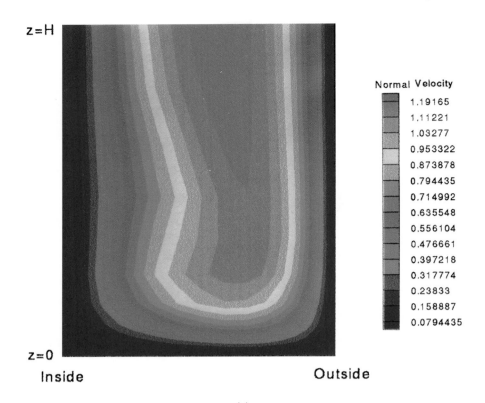

(a)

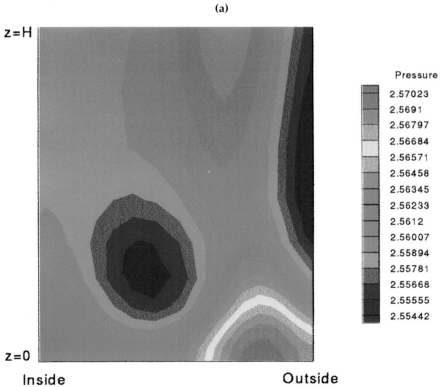

(b)

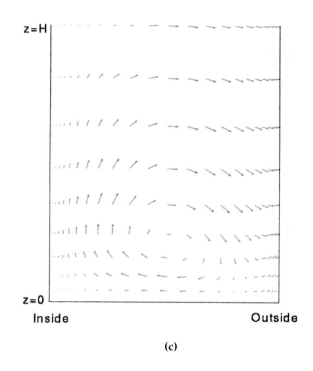

Figure 9 (a) Cross-sectional velocity distribution in the C-plane; (b) cross-sectional pressure distribution in the C-plane; (c) cross-sectional view of the secondary flow in the C-plane.

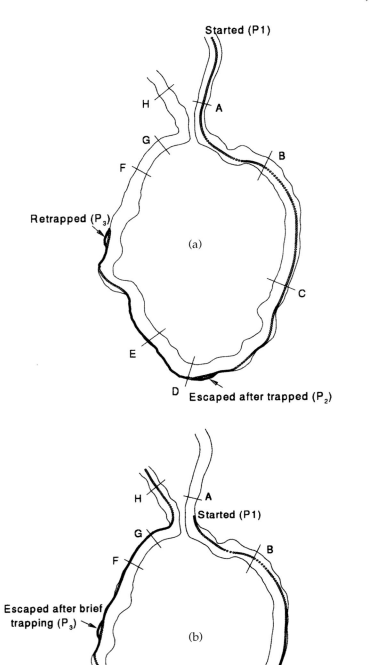

Figure 10 Different trajectories of a receptacle.

in the Far East Asian countries. We were thus able to reconstruct some of the physical flows our ancestors had witnessed, reviving even some amusement and lyricism they had enjoyed during the Curved Water Feast.

Acknowledgments

The authors are grateful to professor Yasuki Nakayama, Future Technology Research Institute, Tokyo, for the literature and photographs, as well as for the discussions he kindly allowed at his home, regarding the Curved water channels in Japan. Appreciation is also expressed to Professor Naixing Chen, Chinese Academy of Sciences, Beiging, for the valuable related materials. Finally, we thank professor Inyong Ham, Pennsylvania State University, for his continued interest and encouragement in this research.

References

Chang K.S., A study of Vessel-Floating Curved Water Passage of Posuk Pavilion in Kyungju from the Fluid Science View Point, *Misoolcharyo* (美術資料), Vol. 46, pp. 101–110 and pp. 114–115, December 1990, National Central Museum, Korea (in Korean).
Issa R., Solution of the Implicitly Discretized Fluid Flow Equations by Operator Splitting, *J. Comp. Phys.*, vol. 62, pp. 40–65, 1986.
Nakayama Y., Aoki K., Oki M., Koksui-no-en Considered from Fluid Mechanics Point of View, *J. Visualization Soc. Jpn.*, Vol. 14, Suppl. 1, pp. 191–194, 1994 (in Japanese).
Shim E.B., Chang K.S., Three-Dimensional Vortex Flow past a Tilting Disc Valve Using a Segregated Finite Element Scheme, *Computational Fluid Dynamics J.*, Vol. 3, No. 1, pp. 205–222, April 1994.
Takeshima T., *The Studies on Ying Tsao Fa Shih*, Vol. 1, Chap. 14, pp. 196–205, JuoKoron Bijiuts Publishing, Tokyo, 1970 (in Japanese).
Zhao Luo, The first writing of a gossip column "Liu Bei Ting", *Beijing Daily Newspaper*, May 22, 1992 (in Chinese).

Index

A

Air bubble burst, turbulent flow, 110–111
Air flow, see Jet flow; Wind entries
Airfoil, leading edge flow, 76–77
Air-water interface
    slopes and elevations, 1–20
        applications, 12–20
        color palette generation, 8–10
        data analysis, 11–12
        methods, 1–3
        optical principle, 3–5
        system design, 5–8
    vorticity, 111
Angiography, 218–219
Aperture, in particle-image velocimetry, 117
Arithmetic-logic unit, 133–134
Artificial intelligence, 216
Asphalt-paved highway, 225–238
    general features, 227–229
    horizontal and vertical patterns, 231–238
    numerical simulation, 226
    potential temperature, 229
    turbulence of wind velocity, 229–231
Asymmetrical circular cylinder, 55–65, 66
Asymmetrical flow, vortex shedding, 58
Asynchronous color particle image velocimetry, 90–97
Attached flows, 101–107, 154–155
Attached symmetrical recirculating zone, 51–52, 53, 65
Attachment points, 45, 46
Autocorrelation, in particle-image velocimetry, 118
Axisymmetric mathematical model, 183–184, 193–195
Azimuthal instabilities, 195

B

Back-illuminated method, 206–208
Baroclinic torque, 108
Bifurcation process, 50, 65
Binarization, 133
Bit-plane images, 134–135
Blade-element/momentum theory, 127
Blood flow, 218–220
Blunt cylinder, turbulent flow, 101–107
Blunt flat plate, turbulent flow, 101–107
Body, see Human body visualization
Body junction, vortex-shedding process, 65, 66
Body walls, see Walls
Bomb, nuclear, 107–110
Boundary conditions, 104
    leading edge flow, 76–77
    low speed air flow, 72–74
    propane jet diffusion flames, 183–184
    thermal plumes over highways, 226
Brain, 216, 217
Bubble burst, turbulent flow, 110–111
Buoyancy, 188–189, 191–192

C

Calibration curves, color particle image velocimetry, 89
Camera distance, 120
Cameras, 200, see also CCD system
Capillary gravity and waves, 13–14
Cardiovascular system, 214, 216, 217, 218–220
Car model, 92–97
Cast iron cylinder, 209
Cavity flow, 136
CCD (charge coupled device) system, 4, 119, 120, 200
    advantages, 40
    particle image velocimetry, 70
Centers (foci), 45, 46, 52, 65
Cervical vertebrae, 215
Charge coupled device system, see CCD systems
Chemisorption visualization, 143–151
China
    Curved Water Feast, 242
    Dragon Washbasin Phenomena, 173–179
Cine camera, 200
Circular cylinder, see Cylinder
Cleaning systems, 148, 149
Coherence lengths, 190
Cold jets, 181–197
    mathematical model, 183–184
    reactive-Mie-scattering technique, 182
    vortical structures
        cross sections, 195, 196
        high-speed flows, 192–195
        low-speed flows, 184–192

Color-coded particle-tracking velocimetry
　image grabbing and coding, 132–134
　posterior image processing, 134–136
　stereo imaging, 138–141
　two-dimensional flows, 136–137
Color mapping technique, free surface slopes
　　　and elevations, 1–20
　applications, 12–20
　data analysis, 11–12
　optical principle, 3–5
　palette generation, 8–10
　system design, 5–8
Color particle image velocimetry, 83–97
　asynchronous, 90–97
　benefits and limitations, 96–97
　synchronous, 84–90
Combustion
　internal combustion engine, 199–211
　　diesel combustion process, 204–208
　　fuel-air mixture preparation, 203–204
　　in-cylinder flow, 201–203
　　lubrication around pistons, 208–211
　　methodology, 200
　　spray, 200–201
　propane jet diffusion flames, 181–197
　　mathematical model, 183–184
　　methodology, 182
　　statistical vs. dynamic studies, 181–182
　　vortical structures
　　　cross sections, 195, 196
　　　high-speed flows, 192–195
　　　low-speed flows, 184–192
Computational simulations, see Numerical
　　　simulations
Computed tomography (CT)
　human body, 214–223
　internal combustion engines, 201
　laser light sheets
　　dynamic, 27–30
　　movable plane, 30–36
　　parameters, 23–27
　　plane dynamic, 27–30
　　simultaneous multiple plane, 36–40
Confinement effect, 54
Convection, see Heat transfer
Copper oleate, 204
Correlation, in particle-image velocimetry,
　　　118–120
Coumarin-6, 210
Critical points, 44, 45, 46
Cross-correlation, in particle-image
　　　velocimetry, 120
Cross-coupling of colors, 11–12
CT, see Computed tomography
Cube, 141

Curved Water Channel and Feast, 241–255
　computational simulation, 245–247
　history, 241–245
　Posuk-Chung Channel, 247–248
　trajectory of receptacle, 248–255
Cylinder
　heat transfer
　　stream experiments, 154–155
　　wind tunnel experiments, 155–167
　turbulent flow, 101–107
　vortex-shedding process, 47–48, 55–65, 66

D

Dean number, 34
Delta wing, 33, 77–81
Diaphragm cell, 148
Diesel combustion, 204–208 Diffusion, 190–191
Digital imaging
　chemisorption visualization, 145–147
　color-coded particle-tracking velocimetry,
　　　132–134
　particle image velocimetry, 172–173
Digital particle image velocimetry, see Particle
　　　image velocimetry
DOLFINS (drastically operator-lightened
　　　finite-element implicit Navier-Stokes)
　　　code, 246–247
Doppler methods, 116, 117, 219
Drag coefficient, 249
Dragon Washbasin Phenomena
　flow and vibration frequency, 173–179
　mechanism, 169–170
　methodology, 171–173
Dyes, 101
Dynamical range of velocity, 88, 90–97
Dynamic stall, 76–77

E

Electrochemical dye visualization, 44
Ellipse, vortex shedding, 63–64
End plate-body junction flow, 52–53, 65, 66
Engine, see Internal combustion engine
Errors
　particle-image velocimetry, 121, 122–123
　random, 121
　systematic, 121

F

Film speed, 120
Fish Washbasin Phenomena, 169–179
Flame-flickering, 188
Flames, see Combustion

Flat plate, turbulent flow, 101–107
Flowing-Receptacle Curved Water, 242
Flow instability, junction flow, 65
Flow mechanisms, see also Jet flow;
    Water flow; Wind entries
  chemisorption of mass transfer, 143–151
  in cube, 141
  particle-streak visualization, 43–67
    applications, 43–45
    basic concepts, 45–48
    nominally 2D bluff body, 55–64
    techniques, 53–55
    vortex-shedding process, 48–53
    wall junction flow, 65, 66
  in square duct, 138–141
  vascular system, 218–220
Flow separation, see Separated flow
Flow time development, 54
Flow visualization, see also specific
    techniques
  applications, 100
  turbulent structures, 99–112
  wind flow, 123–124
Fluorescein sodium, 172
Fluorescence, 210
  laser-induced fluorescence, 103–107, 203–204
  planer laser-induced fluorescence, 203
F numbers, 6
Foci (centers), 45, 46, 52, 65
Four-dimensional body visualization, 213–223
Four-front-mirror device, 138–139
Fourier law, 156 Fourier transforms, 12, 118–119
Fractal theory, 216
Free-slip critical points, 45
Free surface slopes and elevations, color
    mapping technique, 1–20
  applications, 12–20
  data analysis, 11–12
  optical principle, 3–5
  palettes generation, 8–10
  system design, 5–8
Froude number, 229
Fuel-air mixture in engines, 203–204
Fuzzy theory, 216

G

Galerkin finite element method, 247
Gasoline-air mixture in engines, 203–204
Gaussian beam theory, 25
Gravitational field, 183
Gray scales, 132–133

Grid-generated turbulence, 121, 122, 123
Grid singularity, 248

H

Half-nodes, 45
Half-saddles, 45, 47
Halo function, 119
Harmonic peaks, 118
Heart, 214, 216, 217, 218–220
Heated-thin-foil technique, 156
Heat transfer
  to air from cylinder, 153–167
    methodology, 155–157
    sharp-edged vs. hemispherical-nosed, 157–167
  asphalt-paved highway, 225–238
    general features, 227–229
    horizontal and vertical patterns, 231–238
    numerical simulation, 226
    potential temperature, 229
    turbulence of wind velocity, 229–231
  chemisorption visualization, 143–151
  color-coded particle-tracking velocimetry, 136, 137
  in stream from cylinder, 154–155
Helical vortex system, 123–128
Helmholtz-Kelvin vortex structures, 182
Helmholtz's laws, 101
Hemispherical-nosed cylinder, 157–167
Hepatectomy lobectomy, 222, 223
High-speed flows, vortex structures, 192–195
Highways, see Asphalt-paved highway
Hot wire anemometry, 154–155
Hue, saturation, intensity scheme, 9–12
Human Atlas Database, 216–218
Human body visualization, 213–223
  advantages, 213–214
  blood flow, 218–220, 223
  database, 216–218
  hard tissues, 215–216
  skeletal system movements, 220–221
  soft tissues, 216
  three-and four-dimensional, 214–215
  virtual reality, 221–223

I

Image grabbing and coding, 132–134
Image shifting, in particle-image velocimetry, 117–118
Impacting water drops, turbulent flow, 107–110
Infrared thermography of heat transfer, 153–167

methodology, 155–157
   sharp-edged vs. hemispherical-nosed cylinder, 157–167
Input look-up table, 132–133
Instream saddles, 45, 47
Intensity, color particle image velocimetry, 86–88
Internal combustion engine, 199–211
   diesel combustion process, 204–208
   fuel-air mixture preparation, 203–204
   in-cylinder flow, 201–203
   lubrication around pistons, 208–211
   methodology, 200
   spray, 200–201
Inviscid/viscous interaction, 76

J

Japan
   Curved Water Channels, 242–245
   thermal plumes over highways, 225–238
Jet flow
   cleaning systems, chemisorption visualization, 148, 149
   color mapping, 13
   particle image velocimetry
      low speed flows, 72–75
      supersonic, 70–72
   propane diffusion flames, 181–197
      mathematical model, 183–184
      methodology, 182
      statistical vs. dynamic studies, 181–182
      vortical structures
         cross sections, 195, 196
         high-speed flows, 192–195
         low-speed flows, 184–192
Jet spreading, 185
Junction flow, particle-streak visualization, 52–53, 65, 66

K

Kang Xi, 242
Kármán vortex shedding, 104
Kelvin-Helmholtz vortex structures, 182
KL values, 206, 208
Kogelnik gaussian beam theory, 24
Korea, 241–255

L

Lagrangian equation, 248–249
Large eddy simulation turbulence model, 225–238
Laser disk, 200

Laser Doppler anemometry, 116, 117
Laser-induced fluorescence (LIF)
   internal combustion engines, 203–204
   turbulent vortices, 103–107
Laser tomography, See Computed tomography
Leading-edge flow, 76–77
Leading-edge vortex, 58, 63
LES (large eddy simulation turbulence) model, 225–238
Lighting homogeneity, 27–30
Light intensity, color particle image velocimetry, 86–88
Liver, 222, 223
Local line of separation, 159
Local mass transfer, see Heat transfer
Local mass transfer coefficient, 147–148, 156
Locomotion of body, 220–221
Look-up table, 132–133
Low-speed flows
   color-coded particle-tracking velocimetry, 131–142
   vortex structures, 184–192
Lubrication, 208–211

M

Mach disk, 71
Mach number, 71–72
Magnetic resonance imaging (MRI), 214–223
Mass transfer, see Heat transfer
Mass transfer coefficient, 147–148, 156
Mean square error, particle-image velocimetry, 122–123
Membrane systems, chemisorption visualization, 148, 150
Mirrors
   four-front, 138–139
   rotating, 118
Morphological image processing, 135–136
Movable plane sheets, 31–36
Moving boundary problems, 248
MRI (magnetic resonance imaging), 214–223
Muscles, 221

N

NACA profile
Navier-Stokes equations, 246
Near wake, wind turbine, 124
Neo SPAM model, 226
Nodes, 45
Nominally 2D bluff bodies, particle-streak visualization, 55–64

Noninvasive human body visualization, 213–223
Nonpermeable walls, local mass transfer, 143–151
No-slip critical points, 45
NTSC signals, 6
Nuclear bomb, 107–110
Numerical simulations
 Curved Water Channels, 245–247
 heat transfer over highway, 225, 226
 propane jet diffusion flames, 183–184
Nusselt number, heat transfer to air, 157–167

O

Observer, relative convection speed, 46–47
On-line particle image velocimetry, 70–72
Optical Fourier transform techniques, 119
Optical oscilloscopes, 30
Optical principle, free surface slope measurement, 3–5
Orchid Pavilion, 242
Orchid Pavilion Narration, 242
Orthogonal measurements, 86–88
Oscillating ellipse, 63–64
Oscilloscope, 30
Overshoot phenomenon, 229

P

Panning camera, 118
Parallax effects, 121
Particle diameter, 120
Particle image velocimetry (PIV)
 color, 83–97
  asynchronous, 90–97
  benefits and limitations, 96–97
  synchronous, 84–90
 Dragon Washbasin Phenomena, 169–179
 limitations, 83–84
 vs. particle-tracking velocimetry, 131
 technique, 69, 116–121
 three-dimensional flows, 132
 velocity and vorticity fields, 69–82
  air flow, 72–75
  supersonic flow, 70–72
  water flows, 75–81
Particle mass, 248–251
Particle size measurements, 88–90
Particle-streak visualization
 experimental and numerical data, 44
 reconstruction of streamline patterns, 44
 topological mechanisms, 43–67
  applications, 43–45
  basic concepts, 45–48

 nominally 2D bluff body, 55–64
 techniques, 53–55
 vortex-shedding process, 48–53
 wall junction flow, 65, 66
Particle tracking, in particle image velocimetry, 118
Particle-tracking velocimetry
 color-coded, 131–142
  image grabbing and coding, 132–134
  posterior image processing, 134–136
  stereo imaging, 138–141
  two-dimensional flows, 136–137
 internal combustion engines, 201–202
 vs. particle image velocimetry, 131
 three-dimensional flows, 132 Permeable walls, local mass transfer, 143–151
Perturbation of jet flow, 187, 193–195
Photography, in particle-image velocimetry, 117
Photometry, remission, 145
PISO, 247
Pitching-up airfoil, 76–77
PIV, see Particle image velocimetry
Plane dynamic sheets, 27–30
Planer laser-induced fluorescence, 203
Plate heat exchangers, 149, 150
Pliable Jade Pavilion, 242
Poisson equations, 183
Polarizing wavelength coding tomography, 37, 38
Polymer additive, 58
Posterior image processing, 134–136
Posuk-Chung Channel, 247–255
Prevailing Autumn Pavilion, 242
Propane jet diffusion flames, 181–197
 mathematical model, 183–184
 methodology, 182
 statistical vs. dynamic studies, 181–182
 vortical structures
  cross sections, 195, 196
  high-speed flows, 192–195
  low-speed flows, 184–192
Pulse Doppler method, 219
Pyrometry, 206

Q

Qian Long, 242
QUICKEST numerical scheme, 183

R

Rayleigh scattering, 37, 38
Reactive-Mie-scattering technique, propane jet diffusion flames, 181–197

Reattachment, cylinder flow field, 154–155
Receptacle trajectory, Curved Water Channel, 248–251
Recirculating zone
    turbulence, 105–107
    vortex, 51–52, 53, 65
Rectangular foil, tomography of flow, 38, 39
Reflective mode, free surface slope measurement, 3–6
Refractive mode, free surface slope measurement, 4–6
Region of interest, 215
Relaxation time, 249
Remission photometry, 145
Reynolds number
    color particle image velocimetry, 92
    dye tracers, 101
    heat transfer
        cylinder in stream, 154–155
        cylinder wind tunnel experiments, 157–167
    particle image velocimetry, 70
    particle-streak visualization of vortex shedding, 55, 58–64
    Posuk-Chung Channel water flow, 247–248
    turbine blade, 125
    water flow
        over delta wing, 78
        of pitching-up airfoil, 76
Roads, see Asphalt-paved highway
Rotating mirror, 118
Rotating-translating circular cylinder, 48, 49, 58–64

S

Saddle transposition, 45, 47, 48, 49–50, 53, 55, 58
Scanning beam, particle-image velocimetry, 117
Schlieren techniques, 203
Schmidt number, 101
Secondary flow, 247–248
Seiche wave, 170, 173, 178
Self-excited vibration, Dragon Washbasin Phenomena, 169–179
Separated flow
    local line of separation, 159
    particle-streak visualization, 58, 60
    Posuk-Chung Channel water flow, 247–248
    unsteady, 76–77
    vortices, 101–107
Separation bubble
    cylinder heat transfer, 155, 159–162
    turbulence intensity, 155

Separation points, 45, 46
Separatrices, 45
Shadowgraph method, 206
Sharp-edged cylinder, heat transfer, 157–167
Shear layers, 50, 104
    color mapping, 12–13
    jet diffusion flames, 185–195
    particle image velocimetry, 70–77
    wind turbine wakes, 128
Shock-cell, of underexpanded jet, 70–72
Shock wave oscillations, 30
Signal-to-noise ratio, 119
Simultaneous multiple tomography, 36–39
Sinusoidal wavy channels, 148, 149
Skeleton, 215–216, 220–221
Skin-friction lines, 155
Smoke studies
    streaklines, 101
    turbulent vortices, 102
    wind flow, 123–124
Solar radiation, 225–238
Soot formation zone, 206–208
Spacers in membrane systems, 148, 150
Spanwise currents, 52, 55, 65
Spatial light modulator, 119
Speed ratio, vortex shedding, 58–64
Spray
    Dragon Washbasin Phenomena, 169–179
    internal combustion engines, 200–201
Square duct, flow in, 138–141
Stagnation point, 65
Stamp Receptacle Pavilion, 242
Statistical techniques, 2
Stereo color-coded particle-tracking velocimetry, 138
Stereo photography, 2
Stereoscopic imaging, three-dimensional flows, 132, 138–141
Stoker number, 249–250
Stokes diaphragm cell, 148
Streak photography
    Curved Water Channels, 245–246, 247
    vs. particle-image velocimetry, 116
Streaming Receptacle Pavilion, 242
Streamline patterns
    counterrotating vortices, 49
    critical points, 45–48
Stream studies, see Water flow
Strobe, in particle-image velocimetry, 117
Strouhal number, 187
Supersonic flow, 70–72
Supersonic instabilities, 29, 30
Surface half-saddles, 45

Surgical planning systems, 221–223
Synchronous color particle image velocimetry, 94–90

T

Taylor-Goertler vortices, 148, 149
Temperature
　diesel combustion, 206
　propane jet diffusion flames, 189–192
　thermal plumes over highways, 229, 231–238
Thermal convection, see Heat transfer
Three-dimensional visualization
　heat transfer from cylinder, 160
　human body, 213–223
　jet diffusion flames, 195
　laser tomography, 30–38
　particle image velocimetry, 132
　turbulence, 2
Threshold values, 133
Time-sequential 3D images, 214
Tip-vortex, wind flow, 123–124
Tokyo, 225–238
Topological mechanisms, see Flow mechanisms
Tracer direct injection, 99–112
Tracers, 100–101
Trailing vortex
　particle-streak visualization, 58
　wind turbine, 125
Transitional wall jet, 72–75
Translating-rotating circular cylinder, 48, 49, 58–64
Turbulent flow
　applications of research, 99–100
　blunt cylinder and blunt flat plate, 101–107
　bursting air bubbles, 110–111
　color mapping, 15, 17–20
　cylinder in stream, 154–155
　free surface, 2
　impacting water drop and nuclear bomb, 107–110
　jet flow, 192–195
　methodology, 100–101
　particle-image velocimetry, 121–123
　thermal plumes over highways, 229–238
　vascular system, 218–220
Two-color pyrometry method, 206
Two-dimensional visualization
　flow mechanism, 45
　free surface color mapping, 1–20
　　applications, 12–20
　　data analysis, 11–12
　　optical principle, 3–5
　　palette generation, 8–10
　　system design, 5–8
　laser beam sweeps, 30–31
　particle image velocimetry, 132, 136–137
　particle-tracking velocimetry, 132
　turbulence, 2
　vortex shedding, 55–66
　　body wall fixed, 55–58
　　body wall in motion, 58–64
　　wall junction flow, 65, 66

U

Ultrasonography, 218–220
Underexpanded jet, 70–72
Unsteady separated flows, 76–77

V

Vascular system, 214, 216, 217, 218–220
Vegetation, 226
Velocity
　color particle image velocimetry
　　direction, 90–96
　　orthogonal, 86–88
　convection, 136, 137
　Curved Water Channel, 248–251
　Dragon Washbasin Phenomena, 173
　errors, 121
　jet diffusion flames, 183–184
　particle image velocimetry
　　air flow, 72–75
　　dynamical and angular range, 88, 90–97
　　supersonic flow, 70–72
　　turbulence, 121–123
　　velocity and vorticity fields, 69–82
　　water flow, 75–81
　　wind turbine wakes, 125–128
　square duct fluid flow, 140–141
　thermal plumes over highways, 229–238
Ventricle, 219
Vibration frequency, Dragon Washbasin Phenomena, 173–179
Video tape recorder, 200
Virtual reality, 221–223
Viscosity, 190–191
Viscous diffusion, 47
Viscous/inviscid interaction, 76
Visible Human Project, 218
Vortex breakdown, 77
Vortex coalescence, 48, 50, 58
Vortex lines, 101
Vortex-loop evolution, 47–48
Vortex merging, 186, 188, 191, 195

Vortex ring
  Dragon Washbasin Phenomena, 173
  nuclear bomb, 107–110
Vortex shedding
  particle-streak visualization
    nominally 2D bluff body, 55–65, 66
    topological mechanisms, 48–53
  sequences, 47–48
  turbulence, 104
Vortex splitting, 48, 50–51, 55
Vortex structures
  jet diffusion flames
    boundary conditions, 183–184
    cross sections, 195, 196
    high-speed flows, 192–195
    low-speed flows, 184–192
  turbulence, 99–112
    blunt cylinder and blunt flat plate, 101–107
    bursting air bubbles, 110–111
    flow visualization, 99–112
    impacting water drop and nuclear bomb, 107–110
    methodology, 100–101
    particle-image velocimetry, 121–128
Vortices
  color mapping, 15, 16, 17
  color particle image velocimetry, 95–96
  counterrotating, 48, 49
  critical points, 45–48
  Curved Water Channels, 245–248, 251
  delta wing flow field, 77–81
  dimensional change, 51–52, 53
  double-row structure, 74, 188
  heat transfer from cylinder, 161–162
  particle image velocimetry, 69–82
    air flow, 72–75
    supersonic flow, 70–72
    water flows, 75–81
  relative convection speed of observer, 46–47
  secondary, 80
  sinusoidal wavy channels, 148, 149
  tomography
    air flow around delta wing, 33
    water flow downstream of bend, 33–36
  transposition, 48–50, 55
Voxel man, 218

W

Wake
  car model in water channel, 92–97
  vortex shedding, 57, 58
  wind turbines, 121–128
Walls
  local mass transfer, 143–151
  topological pattern, 55–65, 66
Wang Xizhi, 242
Water-air interface, see Air-water interface
Water drop impact, turbulent flow, 107–110
Water flow
  around car model, 92–97
  around rectangular foil, 38, 39
  Curved Water Channel, 241–255
    computational simulation, 245–247
    history, 241–245
    Posuk-Chung Channel, 247–248
    trajectory of receptacle, 248–255
  downstream of bend, 33–36
  Dragon Washbasin Phenomena, 169–179
  heat transfer from cylinder, 154–155
  particle image velocimetry
    leading edge flow, 76–77
    leading edge vortices, 77–81
    low speed flows, 75–76
Wavelength coding tomography, 37–38, 39
Wavy channels, 148, 149
Weighting functions, 184
Wind, thermal plumes over highways, 229–238
Wind tunnel experiments
  delta wing, 33, 34, 35
  heat transfer from cylinder, 153–167
    methodology, 155–157
    sharp-edged vs. hemispherical-nosed, 157–167
Wind turbines, vortex wakes, 121–128
Wind waves
  color mapping, 1–20
    applications, 12–20
    data analysis, 11–12
    optical principle, 3–5
    palettes generation, 8–10
    system design, 5–8
    measurement techniques, 2–3
Wire-frame method, 219

X

Xenon flash lamp, 200
X-ray absorption rate, 215
X-ray angiography, 218–219
X-ray computed tomography, 214
X-ray photography, 214, 220

Y

Ya Shang Ting, 243
Ying Tsao Fa Shih, 242, 243
Young's fringes, 119